Assessing the Relationship Between Propagule Pressure and Invasion Risk in Ballast Water

Committee on Assessing Numeric Limits for Living Organisms in Ballast Water

Water Science and Technology Board

Division on Earth and Life Studies

NATIONAL RESEARCH COUNCIL
OF THE NATIONAL ACADEMIES

THE NATIONAL ACADEMIES PRESS
Washington, D.C.
www.nap.edu

THE NATIONAL ACADEMIES PRESS 500 Fifth Street, N.W. Washington, DC 20001

NOTICE: The project that is the subject of this report was approved by the Governing Board of the National Research Council, whose members are drawn from the councils of the National Academy of Sciences, the National Academy of Engineering, and the Institute of Medicine. The members of the panel responsible for the report were chosen for their special competences and with regard for appropriate balance.

Support for this study was provided by the EPA and U.S. Coast Guard under contract no. EP-C-09-003, TO#11. Any opinions, findings, conclusions, or recommendations expressed in this publication are those of the author(s) and do not necessarily reflect the views of the organizations or agencies that provided support for the project.

International Standard Book Number-13: 978-0-309-21562-6
International Standard Book Number-10: 0-309-21562-5

Photo on the front cover is courtesy of Gregory M. Ruiz. Photo on the back cover is courtesy of Stephan Gollasch.

Additional copies of this report are available from the National Academies Press, 500 5th Street, N.W., Lockbox 285, Washington, DC 20055; (800) 624-6242 or (202) 334-3313 (in the Washington metropolitan area); Internet, *http://www.nap.edu*.

Printed in the United States of America.

THE NATIONAL ACADEMIES

Advisers to the Nation on Science, Engineering, and Medicine

The **National Academy of Sciences** is a private, nonprofit, self-perpetuating society of distinguished scholars engaged in scientific and engineering research, dedicated to the furtherance of science and technology and to their use for the general welfare. Upon the authority of the charter granted to it by the Congress in 1863, the Academy has a mandate that requires it to advise the federal government on scientific and technical matters. Dr. Ralph J. Cicerone is president of the National Academy of Sciences.

The **National Academy of Engineering** was established in 1964, under the charter of the National Academy of Sciences, as a parallel organization of outstanding engineers. It is autonomous in its administration and in the selection of its members, sharing with the National Academy of Sciences the responsibility for advising the federal government. The National Academy of Engineering also sponsors engineering programs aimed at meeting national needs, encourages education and research, and recognizes the superior achievements of engineers. Dr. Charles M. Vest is president of the National Academy of Engineering.

The **Institute of Medicine** was established in 1970 by the National Academy of Sciences to secure the services of eminent members of appropriate professions in the examination of policy matters pertaining to the health of the public. The Institute acts under the responsibility given to the National Academy of Sciences by its congressional charter to be an adviser to the federal government and, upon its own initiative, to identify issues of medical care, research, and education. Dr. Harvey V. Fineberg is president of the Institute of Medicine.

The **National Research Council** was organized by the National Academy of Sciences in 1916 to associate the broad community of science and technology with the Academy's purposes of furthering knowledge and advising the federal government. Functioning in accordance with general policies determined by the Academy, the Council has become the principal operating agency of both the National Academy of Sciences and the National Academy of Engineering in providing services to the government, the public, and the scientific and engineering communities. The Council is administered jointly by both Academies and the Institute of Medicine. Dr. Ralph J. Cicerone and Dr. Charles M. Vest are chair and vice chair, respectively, of the National Research Council.

www.national-academies.org

Committee on Assessing Numeric Limits for Living Organisms in Ballast Water

JAMES T. CARLTON, *Chair*, Williams College/Mystic Seaport, Mystic, Connecticut
GREGORY M. RUIZ, *Vice-Chair*, Smithsonian Environmental Research Center, Edgewater, Maryland
JAMES E. BYERS, University of Georgia, Athens
ALLEGRA CANGELOSI, Northeast-Midwest Institute, Washington, D.C.
FRED C. DOBBS, Old Dominion University, Norfolk, Virginia
EDWIN D. GROSHOLZ, University of California, Davis
BRIAN LEUNG, McGill University, Montreal, Quebec
HUGH J. MACISAAC, University of Windsor, Ontario
MARJORIE J. WONHAM, Quest University, Squamish, British Columbia

NRC Staff

LAURA J. EHLERS, Study Director
ELLEN DE GUZMAN, Research Associate

Water Science and Technology Board

Preface

The human-mediated introduction of species to regions of the world they could never reach by natural means has had great impacts on the environment, the economy, and society. In the ocean, these invasions have long been mediated by the uptake and subsequent release of ballast water in ocean-going vessels. Increasing world trade and a concomitantly growing global shipping fleet composed of larger and faster vessels, combined with a series of prominent ballast-mediated invasions over the past two decades, have prompted active national and international interest in ballast water management.

Following the invasion of European zebra mussels *(Driessena polymorpha)* in the Great Lakes, the United States Congress passed the Nonindigenous Aquatic Nuisance Prevention and Control Act of 1990 (NANPCA), requiring the United States Coast Guard (USCG) to regulate ballast operations of ships. NANPCA was reauthorized and expanded in 1996 with the passage of the National Invasive Species Act. In 2008, the U.S. Environmental Protection Agency (EPA) entered the ballast water management arena by issuing its first Vessel General Permit, under the authority of the Clean Water Act of 1972.

Both the USCG and EPA ballast management programs are undergoing revisions that focus on setting specific post-treatment discharge standards for ballast water. Forthcoming regulatory deadlines prompted the EPA and the USCG to request the National Research Council's (NRC) Water Science and Technology Board (WSTB) to undertake a study to provide technical advice on the derivation of numeric limits for living organisms in ballast water for the next EPA Vessel General Permit and for USCG programs. The sponsoring agencies asked the NRC to:

1. Evaluate the state of the science of various approaches that assess the risk of establishment of aquatic nonindigenous species given certain concentrations of living organisms in ballast water discharges.

2. Recommend how these approaches can be used by regulatory agencies to best inform risk management decisions on the allowable concentrations of living organisms in discharged ballast water in order to safeguard against the establishment of new aquatic nonindigenous species and to protect and preserve existing indigenous populations of fish, shellfish, and wildlife and other beneficial uses of the nation's waters.

3. Evaluate the risk of successful establishment of new aquatic nonindigenous species associated with a variety of ballast water discharge limits that have been used or suggested by the international community and/or domestic regulatory agencies.

Given the nature of this mandate, this report focuses on inoculum density, which is the basis of proposed discharge standards. Nonetheless, it is but one of many variables that determine whether a species will become a successful invader. The Committee recognized at the outset that any method that attempts to predict invasion outcomes based upon only one of many factors that influence the invasion process is likely to be characterized by a high level of uncertainty.

At the request of the sponsors and given the limited time period for the study, it was not within the Committee's charge to propose numeric discharge standards or to evaluate treatment systems that might be used in the future to achieve any such standards. Finally, the report contains a glossary of terms used.

In developing this report, the Committee benefited greatly from the presentations and input of several individuals, including Henry Lee of EPA's Office of Research and Development, Jim Hanlon and Ryan Albert of EPA's Office of Wastewater Management, Richard Everett and Greg Kirkbride of the U.S. Coast Guard, Andrew Cohen of the Center for Research on Aquatic Bioinvasions, Maurya Falkner of the California State Lands Commission, and John Drake of the University of Georgia. We also thank the stakeholders who took time to share with us their perspectives and thoughts about the many complex issues associated with ballast water and other vector management, including Phyllis Green and Scott Smith, National Park Service; Vic Serveiss, International Joint Commission; Gabriela Chavarria and Tom Cmar, National Resources Defense Council; Doug Schneider, World Shipping Council; Azin Moradhassel, Canadian Shipowners Association; and Caroline Gravel, Shipping Federation of Canada. Finally, the committee thanks Miriam Tepper, Andrew Langridge, Kellina Higgins, and Cassandra Elliott—students at Quest University Canada, whose analysis of 47 papers that studied the relationship between propagule pressure and invasion rate greatly informed the discussion of models in Chapter 4.

Completion of this report would not have been possible without the extraordinary efforts of study director Laura Ehlers, who kept us on task and on point at many potential diversions in the road, especially when shiny baubles began to distract us. The genetic disposition of the Committee to see the light at the end of the tunnel as the oncoming train was well-balanced by Laura's equally genetic talent to sort wheat from chaff. Ellen de Guzman very ably supported meeting logistics and travel arrangements.

This report has been reviewed in draft form by individuals chosen for their diverse perspectives and technical expertise, in accordance with procedures approved by the NRC's Report Review Committee. The purpose of this independent review is to provide candid and critical comments that will assist the institution in making its published report as sound as possible and to ensure that the report meets institutional standards for objectivity, evidence, and responsiveness to the study charge. The review comments and draft manuscript remain confidential to protect the integrity of the deliberative process. We thank the following individuals for their review of this report: Sarah Bailey, Great Lakes Laboratory for Fisheries and Aquatic Sciences, Fisheries and Oceans, Canada; Lisa Drake, U.S. Naval Research Laboratory in Key West, Florida; Gustaaf Hallegraeff, University Tasmania, Australia; Russell Herwig, University of Washington, Seattle; Chad Hewitt, Central Queensland University, Australia; Alex Horne, Alex Horne Associates, El Cerrito, California; Christopher Jerde, University of Notre Dame, Indiana; and Daniel Simberloff, University of Tennesse, Knoxville. Although these reviewers have provided many constructive comments and suggestions, they were not asked to endorse the conclusions and recommendations nor did they see the final draft of the report before its release. The review of this report was overseen by William Chameides, Duke University, who was appointed by the NRC's Report Review Committee and by Judith Weis, Rutgers University, who was appointed by the NRC's Division on Earth and Life Studies. Appointed by the National Research Council, they were responsible for making certain that an independent examination of this report was carried out in accordance with institutional procedures and that all review comments were carefully considered. Responsibility for the final content of this report rests entirely with the authoring committee and institution.

James T. Carlton
Committee Chair

Gregory M. Ruiz
Committee Vice-Chair

Contents

Summary

The uptake and release of ballast water and associated sediments by ships is one of the predominant means by which new aquatic invasive species are introduced around the world. Many of these nonindigenous species have caused extensive environmental, economic, and human health impacts in receiving systems. The prospect for future invasions has inspired world-wide efforts to reduce, if not eliminate, the transport and release of living organisms in ships' ballast water. At the present time, a diverse range of governmental organizations and private interests throughout the world are advancing policy (regulations) and approaches (treatment methods) to reduce ballast-mediated invasions.

Over the past ten years, ballast water exchange to reduce the densities of coastal organisms transferred among global regions has been the most common practice to mitigate ballast-mediated transfers of species. In an effort to go beyond the protectiveness afforded by ballast water exchange, the U.S. Environmental Protection Agency (EPA) and the U.S. Coast Guard (USCG) are developing standards limiting the density of organisms in ballast water discharged to U.S. waters. These agencies requested the National Research Council's (NRC) Water Science and Technology Board (WSTB) to undertake a study to provide technical advice to help inform the derivation of numeric limits for living organisms in ballast water for their regulatory programs.

INTRODUCTION

Each year there are more than 90,000 visits, on average, of commercial ships greater than 300 metric tons to U.S. coastal waters including the Great Lakes, discharging about 196 million metric tons of ballast water. There is significant variation among U.S. ports in terms of the number and frequency of vessel arrivals, ballast volume discharged, and source region of the ballast. The organisms collected from unexchanged ballast water and sediments span orders of magnitude in size, ranging from fishes (30 cm) down to microorganisms (~20 nm). Over 15 animal phyla have been detected in ballast (especially common are mollusks, crustaceans, worms, hydromedusae, and flatworms), as well as algae, seagrasses, viruses, and bacteria. The concentration of organisms present within a ship's ballast water exhibits tremendous temporal and spatial variation because of differences in organism abundance among sources and seasons, particular voyage conditions, characteristics of the ships themselves, and the nature of ballast water management practices, among other factors. Numbers as high as 300 million cysts of toxic dinoflagellates have been detected in a single tank. Thorough ballast water exchange serves to reduce concentrations of live coastal organisms by, on average, 88 to 99 percent. The efficacy of exchange is highest when source ports are freshwater locations and lowest when it involves source ports of high salinity. A final consideration is that much ballast water in the future will be treated to a level beyond what can be accomplished by ballast water exchange, and it is this water that will be the primary target of ballast discharge standards.

Studies on the invasion history of North American waters have involved mining occurrence records from the literature and diverse research programs, rather than an organized field-based monitoring program designed explicitly to detect invasions. Thus, today's knowledge about invasions represents an underestimate of the total number of nonindigenous species that have colonized. Among the best studied freshwater systems in the world are the Laurentian Great Lakes, where more than 180 invaders have been detected and described. At least 55 percent of the nonindigenous species that established populations in the Great Lakes from 1959 onward are attributed to ballast water release. For coastal marine ecosystems, California and western North America have received the most in-depth analyses of aquatic invasions, with over 250 nonindigenous species of invertebrates, algae, and microorganisms having become established in tidal (marine and estuarine) waters of California alone. Of these, only about 10 percent are attributed solely to ballast water as a vector, but greater than 50 percent include ballast water as a possible vector among others. Indeed, ballast water is only one of many potential vectors that now transport marine, estuarine, and freshwater species between continents and oceans; others include vessel biofouling, aquaculture, live bait industries, aquarium and pet industries, the live seafood industry, and the availability of thousands of species on the Internet for unregulated purchase and distribution to the public at large. Despite the challenges that polyvectism implies, the enduring principle of vector management is that limiting (reducing) species transfers decreases invasions. It is for this reason that ballast water (and other vector) management and supervision are critical in protecting and preserving the beneficial uses and the indigenous populations of fish, shellfish, and other wildlife in the nation's waters.

The goal of this NRC report is to inform the regulation of ballast water by helping EPA and the USCG better understand the relationship between the concentration of living organisms in ballast water discharges and the probability of nonindigenous organisms successfully establishing populations in U.S. waters. The NRC was *not* asked to propose specific ballast water discharge limits, as that is a risk management decision, *nor* was it asked to evaluate matters related to the technical and engineering aspects of testing, installing, and using ballast water treatment systems. The NRC committee's statement of task, given below, was to evaluate the risk–release relationship in the context of differing environmental and ecological conditions, including estuarine and freshwater systems as well as the waters of the three-mile territorial sea. With respect to development of allowable concentrations of living organisms in discharged ballast water (inoculum density), the committee was asked to:

1. Evaluate the state of the science of various approaches that assess the risk of establishment of aquatic nonindigenous species given certain concentrations of living organisms in ballast water discharges.

- What are the advantages and disadvantages of the available approaches?
- Identify and discuss the merits and practical utility of other additional approaches of which the National Academy of Sciences is aware.
- How can the various approaches be combined or synthesized to form a model or otherwise more powerful approach?
- What are the data gaps or other shortcomings of the various approaches and how can they be addressed within the near and long term?
- Can a "natural invasion rate" (invasion rates based on historic invasion rates), or other "natural" baselines, be reliably established, and if so, how? What utility might such baselines have in informing EPA's derivation of allowable numeric limits for living organisms in ballast water discharges? Can such baselines be established on a national basis, or would this need to be done on a regional or ecosystem basis?

2. Recommend how these approaches can be used by regulatory agencies to best inform risk management decisions on the allowable concentrations of living organisms in discharged ballast water in order to safeguard against the establishment of new aquatic nonindigenous species and to protect and preserve existing indigenous populations of fish, shellfish, and wildlife and other beneficial uses of the nation's waters.

3. Evaluate the risk of successful establishment of new aquatic nonindigenous species associated with a variety of ballast water discharge limits that have been used or suggested by the international community and/or domestic regulatory agencies.

With regard to Task 3, as discussed briefly below and in greater detail throughout the report, the available methods for determining a numeric discharge standard for ballast water are limited by a profound lack of data and information to develop and validate models of the risk–release relationship. Therefore, it was not possible with any certainty to determine the risk of nonindigenous species establishment under existing discharge limits. To address Tasks 1 and 2, this report outlines a process of model development and data collection critical to informing future numeric ballast water discharge standards.

POLICY CONTEXT FOR REGULATING ORGANISMS IN BALLAST DISCHARGE

Chapter 2 discusses the regulatory context surrounding ballast water management, including state, federal, and international guidelines and regulations that are the foundation for the current ballast discharge standards. The two main federal programs for regulating ballast water in the United States are the EPA's Vessel General Permit under the Clean Water Act and the U.S. Coast Guard's authority under the National Invasive Species Act (NISA). Given the prominent role of two independent agencies, as well as 20 years of federal and state legislative activity, a complex network of regulatory arrangements has been created around ballast water discharge. Internationally, the regulatory arrangement is simpler, as exemplified by the Inter-national Maritime Organization's (IMO) *International Convention for the Control and Management of Ship's Ballast Water and Sediments*. The Convention identifies two key standards; D-1 is a ballast water exchange standard, and the D-2 performance standard sets maximum permissible limits on live organisms in ballast effluent based on the size or taxonomic category of organisms.

The statutes that guide the EPA and USCG regulatory programs appear to provide the essential considerations and scope to successfully implement scientifically based numeric standards, given sources of variation (described in Chapter 3). For example, the CWA is capable of addressing place-specific issues, such as characteristics of the receiving system, through the State certification process. Meanwhile, the NISA program is comprehensive with respect to the ship-related modes of introduction. Both statutes allow the implementing agency to be sensitive to critical risk factors such as voyage patterns and frequencies through variable enforcement intensity. Specifically, rigorous enforcement of reporting and implementation actions by industry, incorporating data gathering on living organisms density and diversity in discharge, will greatly facilitate better understanding of the relationship between propagule pressure[1] (in terms of inoculum abundance and density) and the probability of invasion.

SOURCES OF VARIATION INFLUENCING THE PROBABILITY OF INVASION AND ESTABLISHMENT

An assumption in the development of a numeric standard for live organisms per unit volume ballast water discharged is that there is a direct and quantifiable

[1] Propagule pressure is a general term expressing the quantity, quality, and frequency with which propagules are introduced to a given location. Propagules are any living biological material (such as particles, cells, spores, eggs, larvae, and mature organisms) transported from one location to another.

relationship between the density of individuals released in a ballast discharge and the probability of their eventual establishment. While a relationship between inoculum density and establishment probability may exist, many other factors also affect establishment success in aquatic systems, as discussed in Chapter 3. These include the identity (taxonomic composition), sources, and history of the propagules; their frequency of delivery; and their quality. Further influencing the outcome of propagule release is a host of factors that include both species traits and the recipient region's environmental traits. Given the significant differences between source regions; the diversity, abundance, and density of entrained organisms; and the compatibility of source and recipient regions, the prospect of developing a ballast water standard that can be applied to all ships and yield a desired result is daunting.

It is abundantly clear that significantly reducing propagule pressure will reduce the probability of invasions, when controlling for all other variables. There is both strong theoretical and empirical support for this, across a diverse range of habitats, geographic regions, and types of organisms. However, the *precise* nature of the response can vary enormously over species, time, and environments. In short, while inoculum density is a key component, it is but one of scores of variables that can and do influence invasion outcome. **Thus, any method that attempts to predict invasion outcomes based upon only one factor in the multi-dimensional world of the invasion process is likely to suffer from a high level of uncertainty.**

RELATIONSHIP BETWEEN ESTABLISHMENT RISK AND PROPAGULE PRESSURE

Chapter 4 presents the theory underlying the relationship between ballast water organism concentration and the risk (probability) of nonindigenous species establishment (the risk–release relationship), which is focused on the role of propagule pressure. It evaluates the mathematical models that have been developed to express this relationship, discussing their data needs as well as other strengths and weaknesses. The models range from descriptive models that simply represent the shape of the relationship to mechanistic models that define the processes generating the relationship. The models are also distinguished by whether they consider a single species vs. multiple species. Three of the methods discussed are the reaction-diffusion approach, the population viability analysis, and the per capita invasion probability approach—previously reviewed by EPA for their prospects in helping to set a numeric ballast discharge standard. In principle, a well-supported model of the relationship between invasion risk and organism release could be used to inform a ballast water discharge standard. For a given discharge standard, the corresponding invasion risk could be predicted, or, for a given target invasion risk, the corresponding target release level could be obtained.

Ballast water discharge standards should be based on models and be explicitly expressed in an adaptive framework to allow the models to be updated in the future with new information. Before being applied, it is essential that candidate models be tested and compared, and their compounded uncertainty be explicitly analyzed. Only a handful of quantitative analyses of invasion risk–release relationships thus far have tested multiple models and quantified uncertainty.

In the short term, mechanistic *single-species* models are recommended to examine risk-release relationships *for best case (for invasion)-scenario species*. Candidate best-case-scenario species should be those with life histories that would favor establishment with the smallest inoculum density, including fast-growth, parthenogenetic or other asexual reproductive abilities, and lecithotrophic larvae. The greatest challenge in this approach will be converting the results of small-scale studies to an operational discharge standard.

Developing a robust *statistical model* of the risk-release relationship is recommended. It is anticipated that this approach will be more fruitful at a local scale than a nationwide scale. Within a region, this relationship should be estimated across multiple time intervals, rather than from a single point. The effect of temporal bin sizes on the shape of the relationship should be examined. Currently, the greatest challenge in this approach is the insufficient scope and scale of the available data. Since long-term historical data on ballast-organism density are limited, the committee recommends an extremely careful analysis and validation of any proxy variables. There is no evidence that any proxy variable used thus far is a reliable stand-in for organism density.

Models of any kind are only as informative as their input data. In the case of ballast water, both invasion risk and organism density discharged from ballast are characterized by considerable and largely unquantified uncertainty. At the multi-species scale in particular, the existing data (historical invasion records vs. recent ballast surveys) are substantially mismatched in time, and patchy in time, space, and taxonomy; current statistical relationships with these or proxy variables are of dubious value.

OTHER APPROACHES TO SETTING A BALLAST WATER DISCHARGE STANDARD

In the absence of data and models necessary to support a science-based quantitative approach to setting ballast water discharge standards, expert opinion has been a common alternative. Chapter 5 discusses the strengths and weaknesses of non-quantitative, expert opinion-based methods for setting ballast water discharge standards, including two previously reviewed by EPA (the zero-detectable discharge standard and the natural invasion rate approach).

Regarding standards for living organisms in ballast water discharge, expert opinion processes have taken a number of forms and produced a wide range of outcomes. While each expert opinion process discussed has conceptual merit, each is compromised by assumptions, data limitations, or operational difficulties. Despite these drawbacks, expert opinion has resulted in various standards that could provide a manageable baseline for developing scientific models and serve to reduce propagule supply. For example, as ships attempt to meet the IMO D-2 standard, high density discharges and much of the variation in densities of live organisms in ballast discharge will be modulated. This standard provides a starting point for the regulatory process and can facilitate progression to a scientific model.

THE PATH FORWARD

Approaches to setting ballast discharge standards have relied primarily on expert opinion to evaluate the risk–release relationship. The associated history and process have resulted in an array of different international, national, and state discharge guidelines and regulations that seek to reduce propagule supply below that of untreated ballast water. These differences result from both uncertainty about the risk–release relationship and from the diverse approaches of different decision makers and stakeholders. Of the more scientifically based approaches suggested to date and reviewed in this report, descriptive statistical modeling with proxy variables (such as the per capita invasion probability method) is currently available to empirically examine the risk–release relationship because there are data available for ballast volume (derived indirectly from vessel arrival data) and historical invasion rates across estuaries. However, it must be cautioned that these are extraordinarily coarse-level data because (a) vessel arrival data often do not directly translate into a measure of ballast water actually discharged, (b) when actual ballast volume data are available, these do not translate well into known propagule supply and, (c) there is no significant relationship between ballast volume and invasions. **Thus, while statistical modeling has been applied to current datasets, the data are not sufficient in present form to characterize a biologically meaningful relationship, much less estimate the associated uncertainty, to be able to identify with confidence the invasion probabilities associated with particular discharge standards.**

Several actions are needed to advance a robust understanding of the risk–release relationship in order to inform future decisions about ballast water discharge standards. As a logical first step, **a benchmark discharge standard should be established that clearly reduces concentrations of coastal organisms below current levels resulting from ballast water exchange (such as the IMO D-2 standard).** This will serve to reduce the likelihood of invasion in coastal ecosystems beyond that of the present time.

Following the setting of an initial benchmark, a risk–release model or models should be selected as the foundation for the data gathering and analysis effort. One or several of the models described in Chapter 4 should be pursued in the months and years ahead. What model or models is ultimately chosen will reflect the available resources, in terms of time, data, and personnel. At the present time, none of the available models has been validated, due mainly to a lack of key data. Using multiple models with the same data could be valuable to test for concordance. This would also allow one to assess the range of outcomes that would result from proposed ballast water discharge standards. Furthermore, there is considerable worth in transitioning from simple, statistical single-species models to more complicated, multi-species mechanistic models as more data become available.

Finally, a two-track approach should be pursued to obtain both experimental and field-based (descriptive) data. Important early steps should be taken to develop sampling protocols, standardize methods and analytical processes, and create the framework necessary to produce high-quality data specifically needed to populate risk-release models. Experiments can be used to evaluate this relationship and should deliver results over the next three to five years. Field-based descriptive data should also be collected and analyzed to parameterize the same types of models, providing real-world validation of experimental data. Results from such field efforts would be expected to materialize in about ten years.

Recommendations for Experiments

Experiments should be used to estimate the effect of propagule pressure on establishment success, using statistical and probabilistic models. The experiments should (a) be conducted in large-scale mesocosms designed specifically to simulate field conditions, (b) include a diverse range of taxa, encompassing different life-histories and species from known source regions of potential invasions, and (c) include different types of environments (e.g., freshwater, estuarine, and marine water) where ballast discharge may occur.

Initial experimental efforts should focus on single-species risk–release relationships. Ideally, these would include taxa and conditions that are selected as "best-case" scenarios, seeking to maximize invasion success and provide a conservative estimate of invasion probability. Thus, rather than experiments that examine complex and interactive effects of many different environmental and biological variables, a premium is placed on relatively simple initial experiments that provide a significant amount of data across "model" taxa and conditions in a short amount of time. This approach should be applied to multiple species, and serious consideration should be given to selecting the appropriate organisms and conditions.

The experiments should be advanced aggressively, in a directed fashion, to yield results in a three- to five-year time horizon. While this represents a significant investment in effort and resources, it is the more cost- and time-efficient path to obtaining critical data needed to parameterize risk–release models compared to field-based measures. Experiments could potentially identify a solid interim basis for discharge standards, noting the inherent challenges in working with a limited number of species and the assumptions that these would be representative of a broad array of potential invasions. Importantly, these data may also have direct application to other vectors, in addition to ballast water, as they test basic questions about establishment that are relevant to propagule pressure arising from all vectors.

Recommendations for Descriptive Studies

In addition to experiments, descriptive field-based measures are recommended to ground-truth the models, providing a critical validation step to confirm that (a) risk–release relationships are consistent with experimental results and (b) observed invasion rates are consistent with these predictions. Implementing such an effort at one location is not sufficient. This should occur at selected sentinel estuaries (e.g., San Francisco Bay, Chesapeake Bay, and Tampa Bay), chosen to include different coasts, ship traffic patterns, source regions, and environmental conditions. For each sentinel estuary, measures of propagule supply (in ships' ballast) and invasion rate would be made repeatedly over a minimum of a ten-year time horizon to provide a data set for independent analysis and validation of experimental results.

The specific design of data collection needs to be defined explicitly, considering the model(s) being used and making sure that the output will represent the risk–release relationship and directly translate to a discharge standard. While it may be reasonable to explore potential proxy variables as one component, it is critical to not focus extensively on proxies or other variables that may not represent the risk–release relationship. Also critical is an *a priori* estimate of the uncertainty explicit at all scales, as well as sampling effort (number and frequency of measures), in order to properly design measures and interpret and compare predictions. The same data could be used for statistical and probabilistic models, moving toward increasing resolution (e.g., hierarchical probability models described in Chapter 4) if and as appropriate data are available. A comprehensive model would require sampling many vessels (stratified by vessel type, season, and source) and quantifying the concentration of each species present in discharged ballast (as well as volume per discharge event). Field surveys to detect invasions of these species would also need to be conducted coincident with ballast measures.

One possible strategy would be to focus on a subset of target species discharged in ballast water to multiple estuaries. This would considerably reduce the effort required for analysis of ballast water, compared to characterizing the entire community. It may also serve to reduce the sampling effort, and increase the probability of detection, of the target species in field surveys. Intuitively, it would make sense to focus particular attention not only on species that can be identified and counted in ballast samples, but also on species that are unlikely to be polyvectic (such as copepods and mysids), providing the clearest signal for analysis of risk–release relationships associated with ballast water. With this strategy, selection of taxa is critical and should take into consideration biological and environmental requirements (especially whether suitable conditions exist in the specific estuary). A challenge is how generally representative any such species would be. Nonetheless, this would result in single-species models (for multiple species) in parallel to the experimental approach outlined previously.

While collection of field-based descriptive data required for a meaningful analysis of the risk–release relationship is somewhat daunting in scope, recent developments make this more feasible than in the past. First, pending international and national regulations require commercial vessels to install sampling ports that provide representative and standardized samples of ballast discharge. This will provide an important platform for ready access and standardized, comparable samples across vessels and locations. Second, the implementation of ballast water treatment systems will reduce the concentrations, and possibly the diversity, of organisms in ballast water. This may serve to simplify sampling, having less biological material to process for quantitative analysis. Third, the use of molecular genetic tools has dramatically expanded the capacity (and reduced the time, effort, and cost) to detect species, based on DNA. It is perhaps useful to point out that field-based measures outlined above would also serve a broader range of applications, such as providing critical feedback for adaptive management, identifying performance of discharge standards to reduce invasions, and tracking invasions from other vectors concurrently.

***

To date, there has been no concerted effort to collect and integrate the data necessary to provide a robust analysis of the risk–release relationship needed to evaluate invasion probability associated with particular ballast water discharge standards. Existing experimental and field data are of very limited scope. There is currently no program in place to implement either ship-based ballast sampling or field surveys to detect new invasions across sites. On-going research provides confidence that this approach is feasible, but it is scattered across sites and usually short-term in nature. Several models exist which can quantify the risk–release relationship, given sufficient data that are now lacking. This report outlines the paths, using multiple methods over different time frames, that could address these data gaps, and thus provide a robust foundation for framing scientifically supportable discharge standards for ballast water.

1
Setting the Invasive Species Management Stage

INTRODUCTION

Human activities have moved hundreds, perhaps thousands, of species around the world's oceans to regions they would never have reached by ocean currents or other natural vectors. Human-mediated mechanisms that have historically bridged the natural barriers of ocean basins and continents include the global transportation of a diverse range of organisms attached to ships' hulls, burrowing into wooden ships, and living on (and sometimes in) commercial oysters (Ruiz et al., 2000a; Carlton, 2007, 2011). While these mechanisms are still in play today, it is widely recognized that the uptake and release of ballast water and associated sediments by ships is now one of the predominant means by which new nonindigenous species are introduced around the world (Carlton, 1985; Carlton and Geller, 1993; Gollasch et al., 2002; Kasyan, 2010). Many of these invasions have caused extensive environmental, economic, and human health impacts (Carlton, 2001). The prospects for future invasions, and especially associated large impacts, have precipitated world-wide efforts to reduce, if not eliminate, the transport and release of living organisms in ballast water.

The desire to manage ballast is not new. In the 1890s workers in New Zealand considered at-sea disposal, chemical treatment, and quarantined on-shore disposal for solid ballast to control plant invasions (Kirk, 1893). In 1918, the International Joint Commission on the Pollution of Boundary Waters took up the matter of the discharge of contaminated ballast water near municipal water intakes in the Great Lakes, again considering chemical treatment (Ferguson, 1932). While concerns about ballast discharge continued to be voiced in subsequent decades, the invasion in the 1980s of Japanese dinoflagellates (causing harmful algal blooms) in Australia (Hallegraeff, 1998) and of zebra and quagga mussels (leading to a plethora of economic and environmental issues) in the United States and Canada (D'Itri, 1997), motivated the United Nations' International Maritime Organization (IMO) to take up the introduction of nonindigenous species due to the release of ships' ballast as a serious marine environmental issue. Concomitantly, a diverse range of governmental organizations and private interests throughout the world have been advancing policy (regulations) and approaches (treatment methods) to reduce ballast-mediated invasions (see Chapter 2).

Approaches to ballast water treatment have evolved over the past 20 years. Initial emphasis has been on ballast water exchange (BWE) to reduce the densities of coastal organisms transferred among global regions (see below). Recent regulations have focused on limiting the density of organisms that are permitted in the discharge of ships' ballast water (IMO, 2004; see also Chapter 2). This post-treatment load is referred to as a "discharge standard." For air and water quality, these discharge standards were constructed to reduce the potential harm of dissolved or particulate matter (as measured in ppt, ppm, or ppb) to human health. For ballast water, discharge standards reflect the potential (or probability) that living nonindigenous organisms, when released by ships, will become successfully established in geographic regions where they do not occur and to cause subsequent harm to the environment or human health.

The magnitude, complexity, and truly global scale of shipping present some challenges in advancing ballast water treatment. Maritime commerce is estimated to carry 90 percent of world trade, traversing the globe and encompassing a wide range of environmental conditions and unique operational constraints (Figure 1-1; Kaluza et al., 2010). Thus, effective treatment options must consider the appropriate scale and diverse operating conditions of the shipping industry.

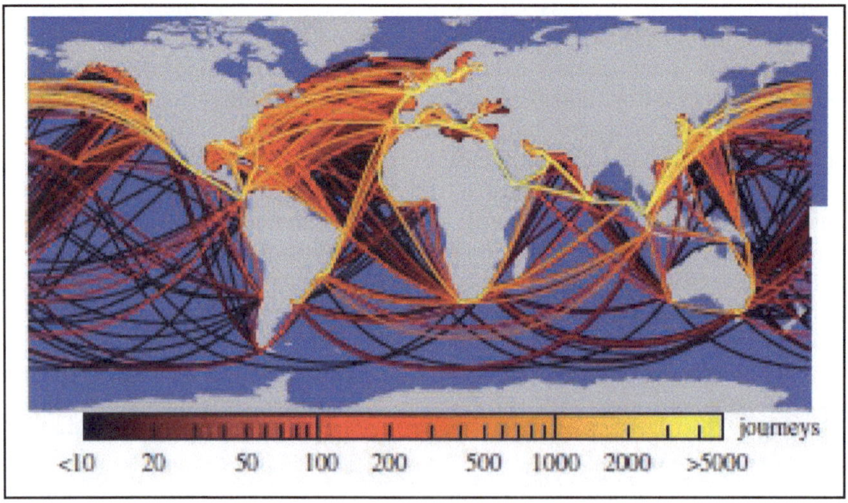

FIGURE 1-1 Routes and ports of the global cargo ship network. Shown are the trajectories of all cargo vessels larger than 10,000 gross tonnage during 2007. The color scale indicates the number of journeys along each route. SOURCE: Reprinted, with permission, from Kaluza et al. (2010). © 2010 by Royal Society Publishing.

THE NUMBER OF VESSELS AND THE VOLUME OF BALLAST WATER IN PLAY

In the United States commercial ships arrive in hundreds of different ports. For commercial ships greater than 300 metric tons, there are over 90,000 arrival events per year to locations in U.S. coastal waters including the Great Lakes (Miller et al., 2010)[2]. Approximately half (48,000) of these are "foreign" or "overseas" arrivals from a last port of call outside the U.S. and Canada. The residual (42,000) are "domestic" or "coastwise" arrivals, which arrive directly from another port within North America. These data are for a two-year period (2006-2007) and highlight the general magnitude of vessel traffic, which exhibits some variation among years.

For this same period, these vessels reported an average annual discharge of 196 million metric tons of ballast water in U.S. coastal waters, based on reports to the National Ballast Water Information Clearinghouse (NBIC; Miller et al., 2010). Ballast water that originated from outside of North America (i.e., that was taken on from a foreign source port) accounted for 28.5 percent of the total discharge volume, and the remainder came from other ports or locations within the U.S. and Canada. These values represent minimum estimates of ballast water discharge, since data were not available for all of the vessel arrivals to U.S. ports.[3]

The number of arrivals and ballast water discharge volume are not evenly distributed among recipient port systems; likewise, the relative contribution of the geographic sources to the number of arrivals and discharge volume is variable (Carlton et al., 1995). The variation in relative importance of different recipient ports is illustrated in Figures 1-2 and 1-3, showing the number of arrivals and volume of ballast discharged across the U.S. that originated from foreign-only sources for 2006-2007, respectively. A similarly high level of spatial variation also exists for "domestic-source" arrivals and ballast discharge in the U.S. (Miller et al., 2010).

While these figures quickly convey the scope of commercial shipping for overseas arrivals, several other key points are highlighted. First, there is tremendous variation among U.S. ports in vessel arrivals and ballast volume. Second, there is a lack of concordance between arrivals and ballast volume for these ports (Figures 1-2 and 1-3). For example, there n are a relatively large number of arrivals in Florida, but this did not translate into a large ballast discharge volume. This reflects the large number of passenger vessels (cruise ships) and container vessels arriving to Florida, and these vessel types routinely

---

[2] This estimate excludes inland traffic (e.g., Mississippi River), and some types of vessels are underrepresented (see Miller et al., 2007, 2010).

[3] An estimated 85% of coastwise arrivals and 86% of foreign arrivals submitted ballast water reports. All vessels are required to submit ballast water reports under U.S. Coast Guard regulations, and there are additional vessels (military and certain commercial ships, such as crude oil tankers engaged in coastwise trade) that are not required to provide this information.

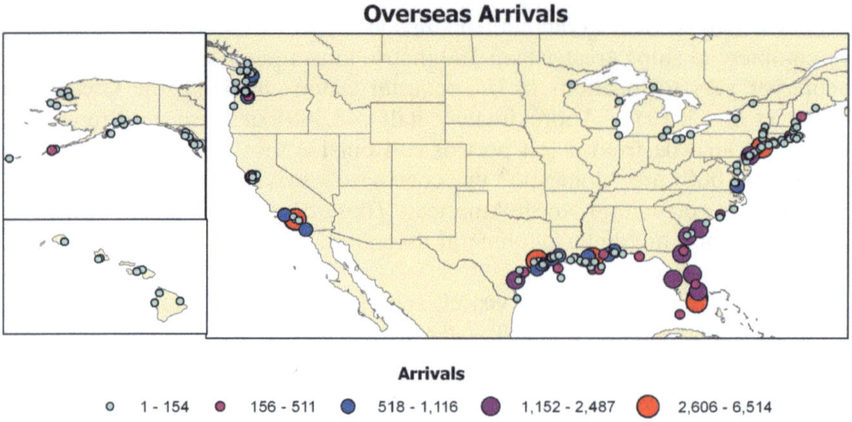

FIGURE 1-2 Number of vessel arrivals to U. S. ports in 2006-2007 that originated from overseas locations (ports of origin outside the U.S. and Canada). SOURCE: Miller et al. (2010).

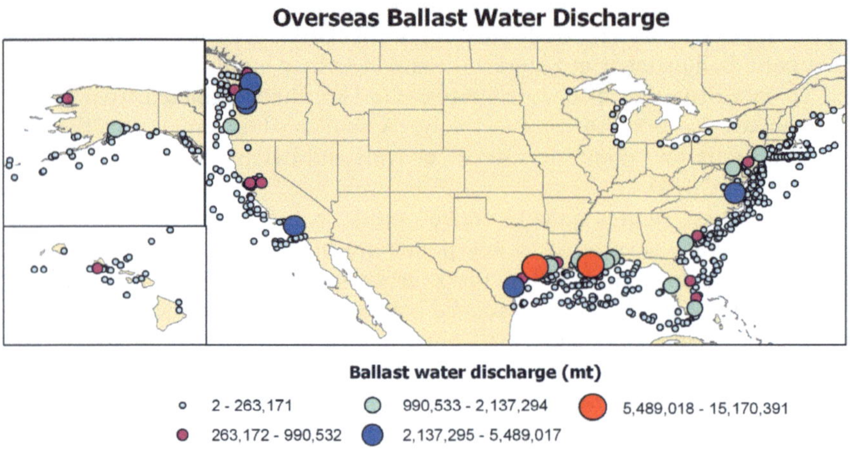

FIGURE 1-3 Amount of ballast water discharged in U. S. ports in metric tons (mt) from overseas sources (outside of U.S. and Canada), regardless of last port of call or route, in 2006-2007. SOURCE: Miller et al. (2010).

discharge very little ballast water. Conversely, the Pacific Northwest and Chesapeake Bay have relatively few arrivals but a large ballast discharge volume, resulting from a large proportion of bulk carriers, which carry and discharge more ballast water than many other vessel types (NBIC; Carlton et al., 1995).

One critical feature that results from these data is that the number of vessel arrivals is not a good predictor or proxy for the volume of ballast water discharged in a port. This is illustrated in Figure 1-4, which shows the total number of arrivals and total ballast water discharged (cumulative across these vessels) for U.S. ports. The weak relationship between these variables has significant consequences for the use of vessel arrivals as an indicator of propagule supply (organisms discharged) via ballast water, as will be discussed later.

Another important dimension of commercial shipping concerns the geographic source of arrivals and the history of ballast water aboard. Figure 1-5 shows the last port of call for foreign arrivals to the U.S. from 2006 to 2007, indicating the relative contribution of different source ports to the arrivals shown in Figure 1-2. This serves to quickly and simply convey the global nature of shipping, which creates connectivity between source and recipient ports, for the transfer of organisms by vessels (associated with ballast water, underwater surfaces, and cargo). Similar projections are available to show the volume of ballast water by source region that is delivered to the U.S. (Miller et al., 2010). The relative contribution of source regions for total ballast to the country differs from that for vessel arrivals, because (1) there are strong differences among ships and routes in the amount of ballast water carried and (2) vessels can simultaneously carry ballast water sourced from multiple ports. Thus, as for recipient ports (Figures 1-2 and 1-3), the number of vessel arrivals may be a poor proxy for relative ballast volume from source regions.

THE DIVERSITY OF ORGANISMS IN BALLAST WATER ENTERING U. S. COASTAL WATERS

Ballast water is typically drawn into tanks from surrounding port water without treatment and thus routinely contains diverse assemblages, from viruses and bacteria to macroinvertebrates (e.g., Carlton, 1985; Carlton and Geller, 1993; Drake et al., 2001). In some cases, organisms as large as medium-sized fish are also drawn into tanks, depending on the size and state of screens used in the sea chest cover or the size of the openings of the gravitation ports (Wonham et al., 2000).

Table 1-1 provides a brief summary of the organisms collected from unexchanged ballast water and sediments arriving in North American coastal waters, which span orders of magnitude in size. The animals collected from ballast water range from fishes (30 cm) down to diapausing eggs (~100 μm); protists, bacteria, and viruses are even smaller in size and more numerous in ballast water.

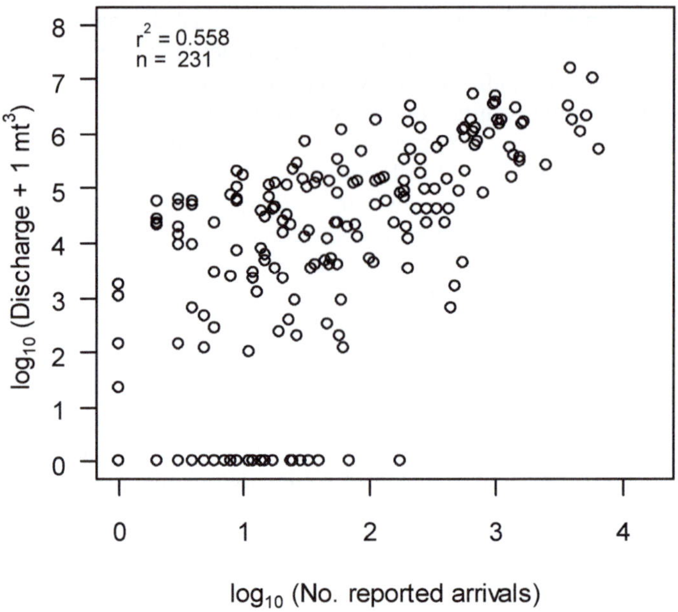

FIGURE 1-4 Relationship between the cumulative number of overseas vessel arrivals and total volume of ballast water discharged for ports in the U.S. for 2006-2007. *mt* = metric ton. SOURCE: National Ballast Information Clearinghouse.

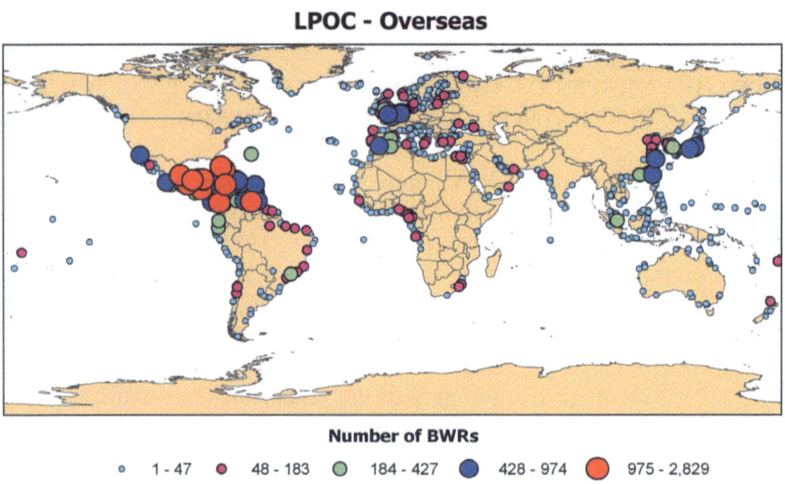

FIGURE 1-5 Last port of call (LPOC) for vessel arrivals to U.S. ports in 2006-2007 that originated from overseas locations. Ballast water aboard ships is not limited to the LPOC as its origin. SOURCE: Miller et al. (2010).

Although this is not meant to represent an exhaustive list, Table 1-1 serves to underscore the taxonomic range associated with ships' ballast tanks that is transported around the globe. Over 15 animal phyla are included (with especially common taxa being mollusks, crustaceans, worms, hydromedusae and flatworms) in addition to algae, seagrasses, viruses, bacteria, and other microorganisms (such as diatoms, dinoflagellates, and other protists; see Table 1-1 for references). Many of the animals in ballast water are in planktonic stages (from larvae to adult), while those in ballast sediments may be adults as well as their diapause (resting) forms.

There are additional analyses of ballast water in other parts of the world that demonstrate the same general picture and add species-level information to studies in North America, but they do not add additional phyla [e.g., Williams et al., 1988 (Australia); Chu et al., 1997 (Hong Kong); Radziejewska et al., 2006 (Russia; sediments); David et al., 2007 (Mediterranean); and Zvyagintsev et al., 2009 (Russia)]. Overall, the size range of ballast-entrained organisms—ranging from 20 nanometers to 30 cm—presents fundamental technical and management challenges in designing control strategies.

Thousands of species may be transported across and between oceans in ballast water on a daily basis (Carlton, 1999), and the cumulative number of species over years to decades is undoubtedly enormous. Most of our understanding comes from studies which have sampled ballast tanks of ships arriving to a discrete location (port) over a period of one to several years, providing only a snapshot of diversity (for a small fraction of discharged ballast water). For example, over 400 species were found in about 150 Japanese wood chip cargo vessels arriving in the Port of Coos Bay, Oregon (Carlton and Geller, 1993). McCarthy and Crowder (2000) reported 342 phytoplankton taxa from only nine ships arriving from a variety of overseas and domestic ports in the Port of Morehead City, North Carolina; one vessel from Europe had over 130 species of diatoms alone. More than 221 species were found in 60 vessels sampled in the Chesapeake Bay (Smith et al., 1999), and 147 species were gathered from 38 ships' samples in the Great Lakes (Duggan et al., 2005). The community composition (species diversity) associated with unexchanged or untreated ballast water will vary tremendously as a function of source, season, and voyage characteristics (e.g., LaVoie et al., 1999; Wonham et al., 2001; Verling et al., 2005). This makes it especially challenging to predict with any confidence the ballast assemblage present in any one ship. In general, most organisms available in the water column and bottom sediments of bays and coastal waters, as well as open-ocean waters, are entrained at some frequency in ballast tanks, unless ships never encounter them or the organisms exceed some size threshold (e.g., marine mammals).

In contrast to what is known about the diversity of metazoans and protists transported by ships' ballast water, very little is known about the corresponding diversity of bacteria and viruses in ballast water. Instead, studies have emphasized their enumeration (e.g., Ruiz et al., 2000; Drake et al., 2001, 2002; Sun et

TABLE 1-1 Diversity of Organisms Collected in Unmanaged Ballast Water and Sediment in Vessels Arriving in North American Coastal and Inland Waters. (The list of micro-organisms is not comprehensive.)

Group	Common name	Coastal	Inland
Animals			
Cnidaria	jellyfish, anemones, hydroids	+	+
Ctenophora	comb jellies (sea gooseberries)	+	
Arthropoda	barnacles	+	+
	copepods	+	+
	decapods (shrimps, crabs, and others)	+	+
	other crustaceans, including euphausids, stomatopods, cumaceans, mysids, isopods, amphipods, ostracods, cladocerans	+	+
	insects		+
	mites	+	+
Tardigrada	water bears		+
Nematoda	thread worms	+	+
Chaetognatha	arrow worms	+	
Mollusca	bivalves (clams, mussels, oysters)	+	+
	gastropods (snails)	+	+
	chitons	+	
Annelida	segmented worms	+	+
Nemertea	ribbon worms	+	
Platyhelminthes	flatworms	+	
Phoronida	horseshoe worms	+	
Bryozoa	bryozoans (moss animals)	+	+
Rotifera	rotifers	+	+
Gastrotricha	gastrotrichs	+	
Echinodermata	sea stars	+	
	brittle stars	+	
	sea urchins	+	
	sea cucumbers	+	
	crinoids		+
Hemichordata	acorn worms	+	
Urochordata	ascidians, larvaceans	+	
Pisces	fishes	+	+
Plants			
Angiosperms	sea grasses	+	
Red and green algae	seaweed	+	

table continues

TABLE 1-1 Continued

Group	Common name	Coastal	Inland
Microorganisms			
Viruses	viruses	+	+
Cyanobacteria	blue green algae	+	+
Other bacteria		+	+
Bacillariophyceae	diatoms	+	+
Dinoflagellata	dinoflagellates	+	+
Ciliophora	ciliated protists	+	
Foraminifera	Forams	+	+
Other "protists"		+	+

Sources for coastal taxa: Smith et al. (1999); McCarthy and Crowder (2000); Ruiz et al. (2000b; Chesapeake Bay); Cordell et al. (2008; Puget Sound); Levings et al. (2004; Vancouver); Carlton and Geller (1993; Coos Bay). Inland taxa: Locke et al. (1993; Great Lakes); Duggan et al. (2005; Great Lakes). Gastrotrich data from Carlton (1985). Sources for microbial taxa: Drake et al. (2001, 2005); Burkholder et al. (2007); Klein et al. (2010); Reid et al. (2007).

al., 2010), sometimes with a taxonomic focus on selected groups (e.g., Ruiz et al., 2000; Drake et al., 2005; Burkholder et al., 2007; Doblin et al. 2007). To the Committee's knowledge, no one has undertaken a metagenomics study (cf. Agogué et al., 2011) of bacteria or viruses in a ballast water context.

ORGANISM CONCENTRATION IN BALLAST WATER

As with community composition, the concentration of organisms present within a ship's ballast water exhibits temporal and spatial variation. This is driven in part by differences in the organism abundances among sources and seasons, but there can also be significant differences in the ballast assemblages of two nearly identical vessels, when sailing from the same port and time period, reflecting the patchy distribution of plankton during ballast operations. Furthermore, even if vessels begin with similar communities, these may diverge through time, as a result of particular voyage conditions and duration or characteristics of the ships themselves (e.g., antifouling coatings). Finally, the nature of any ballast water management practices will influence the concentration of organisms in discharged ballast water.

Past studies provide some estimates of abundances for various types of organisms in ballast tanks. Table 1-2 indicates concentrations of particular organism types found in the ballast water of vessels that were sampled upon arrival to various ports (in the regions indicated). All of these studies were done before ballast water exchange was implemented, providing insight into concentrations

TABLE 1-2 Concentration of Various Organism Types Reported in Ballast Water Sampled Upon Arrival of Ships to Particular Regions

	Concentrations in unexchanged ballast water		
	mean	SE	n
Zooplankton (organisms/m^3)			
Chesapeake Bay (overseas)	1.94×10^3	10×10^1	113
Chesapeake Bay (domestic)	1.75×10^4	1.5×10^3	37
Prince William Sound, Alaska	1.26×10^4	5.53×10^3	169
Phytoplankton (cells/liter)			
Europe	2.99×10^5	1.84×10^5	273
Bacteria (cells/liter)			
Chesapeake Bay	8.3×10^8	1.7×10^8	9
Chesapeake Bay	8.03×10^8	1.88×10^8 (sd)	53
Virus-like Particles (vlp/liter)			
Chesapeake Bay	7.4×10^9	2.3×10^9	7
Chesapeake Bay	1.39×10^{10}	1.57×10^{10} (sd)	31*

SOURCES: Data for Chesapeake Bay are from Minton et al. (2005; zooplankton), Ruiz et al. (2000a), and Drake et al. (2007; bacteria and virus-like particles); data for Alaska are from Hines et al. (2000; zooplankton); data for phytoplankton are from International Maritime Organization (2004). In general, zooplankton refers to organisms collected on nets >50 μm in mesh size, and phytoplankton includes diatoms, dinoflagellates, and other photosynthetic protists.

*It is unclear whether these 31 samples represent exchanged or unexchanged ballast water.

SE = is the standard error, n = number of ballast tanks sampled for concentration estimates.

in unexchanged or untreated ballast water as was common before the mid-1990s. These data may not be representative of certain regions or the country as a whole or predictive of future densities as exchanged and/or treated ballast water becomes more common. Nevertheless, when scaled to the volume of ballast discharged into U.S. waters (196 million metric tons in recent years), these estimates underscore the approximate magnitude of historic biotic transfers due to ships' ballast.

While most past research on organisms in ballast tanks has focused on waterborne assemblages, it is also clear that bottom communities can develop within ballast tanks that can include a diverse range of biota, including adults, larvae, eggs, and resting stages. Very high densities of resting stages can accumulate within ballast tanks. In a survey of 343 vessels in Australia, Hallegraeff and Bolch (1992) found that 65 percent had sediments, all of which contained diatom resting spores. Further, they detected resting stages (cysts) of toxic dinoflagellates in the tanks of 16 vessels and estimated > 300 million cysts of one such species in a single tank. Based on this research and further studies of resting stages for a variety of taxa in the Great Lakes and elsewhere (e.g., McCarthy and Crowder, 2000; Bailey et al., 2005; Fahnenstiel et al., 2009), it is clear that such bottom sediments serve as "seed banks" with viable organisms that can be released from ballast tanks during operations.

Effect of Ballast Water Management

Ships arriving to U.S. waters from overseas are currently required to "manage" their ballast water before discharge into waters of the U.S., and some vessels on coastwise domestic routes are also required to manage their ballast before discharge (Chapter 2). At the current time, ballast water exchange is the only method that is readily available to most vessels. As outlined in Chapter 2, ballast water exchange is being replaced by treatment to specific discharge standards, which are considered more stringent for some organism types (Minton et al., 2005).

In general, ballast water exchange operates to reduce the concentration of coastal organisms that are transferred among global regions, by transferring water from a ship's ballast system to the environment, with concomitant or subsequent uptake of water. Coastal organisms are considered less likely to survive under oceanic conditions, and oceanic organisms are considered less likely to colonize coastal and inland waters, due to habitat and environmental mismatch.

A diverse range of studies, mainly for the greater than 50 μm size class, demonstrate the effect of ballast water exchange on the original contents of ballast tanks. Available data suggest the process of ballast water exchange removes on average 88 to 99 percent of waterborne contents of ballast tanks when performed properly (see review by Ruiz and Reid, 2007). For freshwater and estuarine biota, exposure to high salinity waters often also results in osmotic shock and mortality, further increasing the efficacy of ballast water exchange (Santagata and Ruiz, 2007; Santagata et al., 2008). This combined effect is demonstrated in shipboard experiments, which exposed ballast tanks (either initially ballasted or not ballasted, considered 'no ballast on board' or NOBOB) to saltwater (Gray et al., 2007; Bailey et al., 2011). As shown in Figure 1-6, the mean abundance of freshwater invertebrates, measured as the number of individuals per cubic meter, was significantly reduced (>99.99 percent) following salt-water flushing. In addition, the variation (standard error) was reduced greatly following exchange, further indicating the removal of high density discharges, both for total abundance and freshwater species alone.

While ballast water exchange and salt-water flushing (in the case of NOBOB tanks) have a strong effect in reducing original organisms, including species considered high-risk in the case of the Great Lakes, residual biota are still present in exchanged ballast tanks. For example, Table 1-3 summarizes a substantial amount of empirical data on the abundance of planktonic organisms in ships that have exchanged ballast water prior to entering the Great Lakes. It

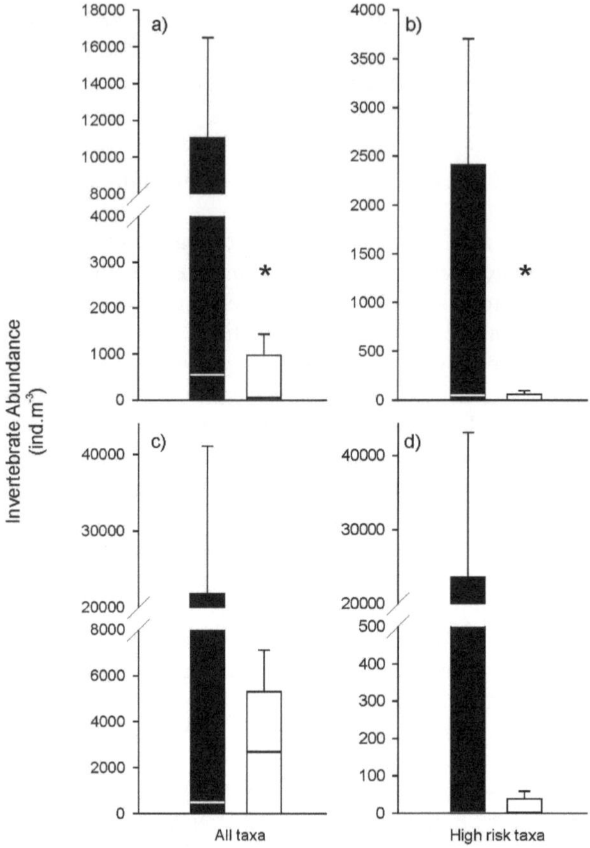

FIGURE 1-6 Mean (+S.E.) abundance of invertebrates recorded from 'no ballast on board' (a and b panels) and ballasted (c and d panels) ships before (black bars) and after (white bars) the introduction of salt-water flushing and ballast water exchange, respectively. Median values are indicated by horizontal white or black lines superimposed on bars. Left panels include data for all taxa; right panels present data only for high risk taxa known to inhabit fresh- or brackish-water habitats. * = P<0.05. SOURCE: Reprinted, with permission, from Bailey et al. (2011). © 2011 by American Chemical Society.

shows that total abundance tends to be slightly higher for invertebrates and dinoflagellates in ballast from coastal vessels, while transoceanic vessels have higher abundances of diatoms and bacteria. For benthic organisms, total abundance was higher in coastal exchanged vessels.

In the case of vessels arriving to the Great Lakes, with tanks having been exposed to seawater by ballast water exchange, many (most) of the waterborne organisms are likely to be marine species instead of freshwater biota that can colonize if discharged. However, many species produce resting stages (e.g., cysts, resting eggs, ephippia) that may survive ballast water exchange and be present in ballast sediments (e.g., Fahnenstiel et al., 2009; Gray and MacIsaac, 2010). Likewise, some diatoms can live for decades or longer in anoxic sediments and are unlikely to be affected by osmotic shock associated with saltwater flushing or ballast water exchange (E. Stoermer, personal communication, October 25, 2010).

Although ballast water exchange serves to reduce the transfer of coastal organisms, there are still residual biota that can colonize coastal recipient ports upon ballast discharge. The efficacy of ballast water exchange is likely lowest when it involves source and recipient ports of high salinity, as there is no added benefit of "osmotic shock" to residual organisms. In such cases, without added mortality, Minton et al. (2005) estimated that vessels implementing ballast water exchange would still deliver a mean of > one million coastal zooplankton alone from the original source per discharge event from the original source. The numbers are greater for smaller organisms, simply because their densities in ballast tanks are much greater than those for zooplankton (Table 1-2).

A final consideration is that much ballast water in the future will be treated, and it is this water that will be the primary target of ballast discharge standards. Though routine treatment by most vessels may be years off (see Chapter 2), early data from land-based testing of ballast treatment systems using ambient source water provide insight into the likely characteristics of treated discharge. First, densities of live organisms can be expected to range within a few orders of magnitude around the discharge standard. Compositionally, rather than a sparser version of the diverse untreated intake assemblage, the post-treatment assemblage may be dominated by species which prove resilient to the specific treatment process and which find the post-treatment environment favorable for survival or re-growth. Treatment by filtration, for example, yields asymmetrical reductions in organisms across taxa largely due to size, but also because of morphology (Cangelosi et al., 2007). Similarly, biocidal processes will affect species differently. Tests of a chlorine treatment system at the Great Ships Initiative land-based Ballast Water Treatment testing facility have shown large shifts in assemblages due to treatment (Cangelosi et al., 2010). Thus, biota in treated water may become much simpler to characterize in terms of variation in concentration and composition, especially as only a limited number of treatment processes will prove effective, safe, and operationally feasible.

TABLE 1-3 Densities of Planktonic Invertebrates (in Water), Benthic Invertebrates (in Sediment), Dinoflagellates, Diatoms, and Bacteria in Ships Arriving to The Great Lakes. Abundances in Water are per 1000 L, Abundances in Sediment are per m^3.

Taxa	Transoceanic exchanged ships		Coastal exchanged ships	
	Total mean abundance (S.E.)	NIS mean abundance (S.E.)	Total mean abundance (S.E.)	NIS mean abundance (S.E.)
Water				
Invertebrates	522.7 (170.4)	421.6 (173.3)	2742.6 (1071.6)	2414.7 (912.3)
Dinoflagellates	2.8×10^4 (1.2×10^4)	2.4×10^4 (1.1×10^4)	6.4×10^4 (3.3×10^4)	6.4×10^4 (3.3×10^4)
Diatoms	1.8×10^6 (1.1×10^6)	8.1×10^4 (4.7×10^4)	1.4×10^5 (8.1×10^4)	9.0×10^4 (7.9×10^4)
Bacteria	7.5×10^{11} (5.6×10^{10})	N/A	8.2×10^{11} (9.2×10^{10})	N/A
Viruses	9.4×10^{12} (1.5×10^{12})	N/A	6.8×10^{12} (1.1×10^{12})	N/A
Sediment				
Invertebrates	8.9×10^5 (2.6×10^5)	1.4×10^4 (1.1×10^4)	1.2×10^6 (2.8×10^5)	1.8×10^3 (1.8×10^3)
Dinoflagellates	6.4×10^4 (1.2×10^4)	6.4×10^4 (1.2×10^4)	9.7×10^4 (2.9×10^4)	9.7×10^4 (2.9×10^4)
Diatoms	3.2×10^9 (1.9×10^9)	9.5×10^6 (5.7×10^6)	N/A	N/A
Bacteria	N/A	N/A	N/A	N/A
Viruses	N/A	N/A	N/A	N/A

N=15, 15, 14, 12, and 12 ships processed for invertebrates, dinoflagellates, diatoms, bacteria, and viruses, respectively, in water of transoceanic ships that exchanged their water ("transoceanic exchanged"). N=13, 6, and 9 ships processed for invertebrates, dinoflagellates, and diatoms, respectively, in sediment of transoceanic ships that exchanged their water ("transoceanic exchanged"). N=4 for water and 5 for sediment for coastal exchanged vessels.

NIS = Nonindigenous Species. N/A = data not available.

Data collected between 2007-2009 inclusive.

SOURCE: Data Courtesy of E. Briski and H. MacIsaac, Canadian Aquatic Invasive Species Network.

U.S. INVASIONS FROM BALLAST WATER

A diverse range of studies have evaluated invasion history of North American waters, examining the patterns in terms of date of first detection, mechanisms of introduction (or vector), native region, and source region (e.g., Carlton, 1979; Mills et al., 1993; Cohen and Carlton, 1995; Ruiz et al., 2000a; Holeck et al., 2004; Wonham and Carlton, 2005; Fofonoff et al., 2009; Kelly et al., 2009). In general, these analyses involved synthesis of occurrence records, which are gleaned from the literature and a diverse range of research programs, instead of an organized field-based research (monitoring) program designed explicitly to detect invasions as they occur. For this reason, it is important to recognize that (1) the resulting knowledge about invasions represents an underestimate of the total number of nonindigenous species that have colonized and (2) only the date of detection is certain, as the lag-time from invasion to detection is unknown (Ruiz et al., 2000a; Solow and Costello, 2004).

The Laurentian Great Lakes are among the best studied freshwater ecosystems in North America, if not the world, with a documented invasion history that dates back to at least 1830 (Mills et al., 1993). More than 180 invaders are now known to be established in the Great Lakes (Ricciardi, 2006). The taxonomic composition of invaders has changed dramatically over time, reflecting changes in different vectors over time. In particular, the switch from solid to liquid ballast in commercial cargo vessels resulted in a wholesale change in nonindigenous species (NIS) composition (Mills et al., 1993). Plants dominated early ship-mediated NIS, while invertebrates and phytoplankton have dominated post-1900 (Holeck et al., 2004). Conservatively, 55 percent of the nonindigenous species that established populations in the Great Lakes during the period following expansion of the St. Lawrence Seaway (from 1959 onward) are attributed to ballast water release (Kelly et al., 2009), although this number could be as high as 70 percent (Holeck et al., 2004).

For coastal marine ecosystems, California and western North America have received the most in-depth analyses of aquatic invasions (Carlton, 1979; Cohen and Carlton, 1995). Over 250 nonindigenous species of invertebrates, algae, and microorganisms (excluding vertebrates and vascular plants) are considered established in tidal (marine and estuarine) waters of California (Ruiz et al., 2011). Of these, only about 10 percent are attributed solely to ballast water as a vector. However, greater than 50 percent include ballast water as a possible vector. This is because many species have life stages and invasion histories that make it possible for the initial introduction to occur by one of several mechanisms, including ballast water, biofouling of vessels' hulls, and transfer of shellfish.

A FURTHER CHALLENGE: THE POLYVECTIC WORLD

The above challenges noted, it is critical to set the role of ballast water and sediment release into the larger vector picture. Ballast water is one of many potential vectors that now transport marine, estuarine, and freshwater species between continents and oceans. These additional vectors form what is known as a *polyvectic* world (Carlton and Ruiz, 2005) and include vessel fouling (on and in many regions of a ship), aquaculture, live bait industries, aquarium and pet industries, the live seafood industry, and the availability of hundreds if not thousands of species on the Internet for unregulated purchase and distribution to the public at large (Lodge et al., 2006). As a result, it is often a challenge to determine which vector or vectors, from a sea of dispersal mechanisms, has led to a particular invasion. Failure to address this multiplicity of vectors with the same intensity and funding that have been applied to ballast water management will result in continued invasions. In short, even when robust and enforced ballast management is achieved, the management community should be prepared for, and not be surprised when, invasions continue.

Coupled with polyvectism is the reality that virtually all management scenarios apply regulatory *filters* rather than complete *barriers* to any vector. A superb example of this is the work of the United States Department of Agriculture's Animal and Plant Health Inspection Service (APHIS), whose responsibilities for the interception of unwanted nonindigenous species date back to 1854 (note that U.S. interests in managing ballast water formally commenced in 1990). Despite the monitoring (inspection, interception, and quarantine) systems in place, and despite the extensive statutory authority wielded by APHIS, new pestiferous insect and plant invasions (for example) occur annually. This is because the holes in the management filtration matrix expand (or contract) over time and space and are at the mercy of the frequency and intensity of inspection, the volume of cargo inspected, human behavior that seeks to circumvent inspection and interception, and yet other factors.

Despite all of the challenges and lacunae outlined above, the enduring value of pursuing vector management is that *control decreases invasions*. In the absence of APHIS and similar federal and state agencies, it is staggering to imagine what the economic, environmental, and societal impacts of terrestrial plant and insect invasions would be in the United States. It is for this reason that ballast water (and other vector) control, restraint, and supervision are critical, and will prove of inestimable value in protecting and preserving the beneficial uses and the indigenous populations of fish, shellfish, and other wildlife in the nation's waters.

REQUEST FOR THE STUDY AND REPORT ROADMAP

There are two main federal programs for regulating ballast water in the United States—EPA's Vessel General Permit under the Clean Water Act and the

U.S. Coast Guard's authority under the National Invasive Species Act (described in detail in Chapter 2). Both programs are undergoing revision and analyses in the near future, which prompted the EPA Office of Wastewater Management and the USCG to request the National Research Council's (NRC) Water Science and Technology Board (WSTB) to undertake a study to provide technical advice to help inform the derivation of numeric limits for living organisms in ballast water for the next Vessel General Permit and for regulatory programs of the USCG.

Both EPA and the USCG desire a federal ballast water management program that will be more effective than ballast water exchange-based requirements in preventing the establishment of new aquatic nonindigenous species through the discharge of ships' ballast water. To improve the regulation of ballast water, the agencies seek to better understand and relate the concentration of living organisms in ballast water discharges (inoculum density) to the probability of nonindigenous organisms successfully establishing populations in U.S. waters. Although the scientific understanding of this relationship is limited, several organizations have created or are in the process of creating numeric standards for ballast water discharges, expressed as limits on the concentrations of living organisms per unit volume.

This report focuses on the initial survival of aquatic nonindigenous species upon release from ballast water and subsequent establishment of a reproducing population, because it is thought that lowering the total concentration of organisms in ballast water is critical to reducing the risk of a successful invasion. Other factors that affect the *overall* successful establishment of nonindigenous species—such as their interface with a transport vector, such as a ship; vector uptake of specific species; survival of the nonindigenous species during transport events; ballast water treatment to reduce NIS numbers; and release of nonindigenous species from the vector—are not the focus. It should be noted that the NRC was *not* asked to propose specific ballast water discharge limits, as that is a risk management decision, *nor* was it asked to evaluate matters related to the technical and engineering aspects of testing, installing, and using ballast water treatment systems on board vessels. The latter topic would include what types of technologies exist and are available for use in the on-board treatment of ballast water discharges, what discharge standards can be reliably achieved by the ballast water treatment systems currently on the market or under development, and what are the technological constraints or other impediments to the development of ballast water treatment technologies. These topics are being considered by a Science Advisory Board committee of the EPA.

The statement of task reads "EPA and the USCG request the NRC to conduct a study that will significantly inform their efforts to derive environmentally protective numeric ballast water discharge limits in the next Vessel General Permit and other programs. The study will take into account estuarine and freshwater systems, including the Great Lakes and other inland navigable waters, as well as the waters of the three-mile territorial sea, considering what implications their differing environmental and ecological conditions might have for the development of allowable concentrations of living organisms in discharged ballast water. Specifics tasks are outlined below.

1. Evaluate the state of the science of various approaches that assess the risk of establishment of aquatic NIS given certain concentrations of living organisms in ballast water discharges.

- What are the advantages and disadvantages of the available approaches?
- Identify and discuss the merits and practical utility of other additional approaches of which the NAS is aware.
- How can the various approaches be combined or synthesized to form a model or otherwise more powerful approach?
- What are the data gaps or other shortcomings of the various approaches and how can they be addressed within the near and long term?
- Can a "natural invasion rate" (invasion rates based on historic invasion rates), or other "natural" baselines, be reliably established, and if so, how? What utility might such baselines have in informing EPA's derivation of allowable numeric limits for living organisms in ballast water discharges? Can such baselines be established on a national basis, or would this need to be done on a regional or ecosystem basis?

2. Recommend how these approaches can be used by regulatory agencies to best inform risk management decisions on the allowable concentrations of living organisms in discharged ballast water in order to safeguard against the establishment of new aquatic NIS and to protect and preserve existing indigenous populations of fish, shellfish, and wildlife and other beneficial uses of the nation's waters.

3. Evaluate the risk of successful establishment of new aquatic NIS associated with a variety of ballast water discharge limits that have been used or suggested by the international community and/or domestic regulatory agencies."

Two documents that summarize an understanding of the risk of invasion for nonindigenous species from ballast water were critical to the work of the committee. Lee et al. (2010) summarized and analyzed seven approaches that have been used or are proposed to either predict the probability of invasion or predict or establish an ecologically "acceptable" concentration of living organisms in ballast water discharges. In addition, in April 2008 the U.S. Coast Guard completed a Draft Programmatic Environmental Impact Statement (DPEIS) accompanying its proposed ballast water discharge standards rulemaking under NANPCA.

Chapter 2 of this report discusses the regulatory context surrounding ballast water management, including state, federal, and international guidelines and regulations that are the foundation for the current ballast discharge standards. Chapter 3 discusses the many sources of variability that ultimately control the rate of invasion from organisms present in ballast water. The prospect of developing a ballast water standard that can be applied to all ships is daunting because ships are coming from all over the world, with significant differences in source regions; in the diversity, abundance, and density of entrained organisms; and in the compatibility of source and recipient regions.

Chapter 4 presents the theory underlying the relationship between ballast water organism concentration and the risk of NIS establishment, which is focused on the role of propagule pressure. It analyzes the mathematical models that have been developed to express this relationship, discussing their data needs and other strengths and weaknesses. In this chapter, three of the methods for setting ballast water standards found in the Lee et al. (2010) report are discussed, including the reaction-diffusion approach, the population viability analysis, and the per capita invasion probability approach. Chapter 5 analyzes and critiques the non-quantitative, expert-opinion-based methods for setting ballast water discharge standards presented in Lee et al. (2010), including the zero-detectable discharge standard and the natural invasion rate approach. The report paves a way forward in Chapter 6 with conclusions and recommendations for setting numeric ballast water discharge standards for the next iteration of the Vessel General Permit and USCG regulations.

REFERENCES

Agogué, H., D. Lamy, P. R. Neal, M. L. Sogin, and G. J. Herndl. 2011. Water mass-specificity of bacterial communities in the North Atlantic revealed by massively parallel sequencing. Molecular Ecology 20:258–274.

Bailey, S. A., I. C. Duggan, P. T. Jenkins, and H. J. MacIsaac. 2005. Invertebrate resting stages in residual ballast sediment of transoceanic ships. Canadian Journal of Fisheries and Aquatic Sciences 62:1090–1103.

Bailey, S. A., M. G. Deneau, L. Jean, C. J. Wiley, B. Leung, and H. J. MacIsaac. 2011. Evaluating efficacy of an environmental policy to prevent biological invasions. Environmental Science and Technology 45:2554–2561.

Burkholder, J. M., G. M. Hallegraeff, G. Melia, A. Cohen, H. A. Bowers, D. W. Oldach, M. W. Parrow, M. J. Sullivan, P. V. Zimba, E. H. Allen, C. A. Kinder, and M. A. Mallin. 2007. Phytoplankton and bacterial assemblages in ballast water of U.S. military ships as a function of port of origin, voyage time, and ocean exchange practices. Harmful Algae 6:486–518.

Cangelosi, A. A., N. L. Mays, M. D. Balcer, E. D. Reavie, D. M. Reid, R. Sturtevant, and X. Gao. 2007. The Response of Zooplankton and Phytoplankton from the North American Great Lakes to Filtration. Harmful Algae 6:547–566.

Cangelosi, A., L. Allinger, M. Balcer, N. Mays, T. Markee, C. Polkinghorne, K. Prihoda, E. Reavie, D. Reid, H. Saillard, T. Schwerdt, H. Schaefer, and M. TenEyck. 2010. Report of the Land-Based Freshwater Testing by the Great Ships Initiative of the Siemens SiCURE™ Ballast Water Management System for Type Approval According to Regulation D-2 and the Relevant IMO Guidelines. Great Ships Initiative.

Available online at: *http://www.nemw.org/GSI/GSI-LB-F-A-1.pdf.* Last accessed June 1, 2011.

Carlton, J. T. 1979. History, biogeography, and ecology of the introduced marine and estuarine invertebrates of the Pacific Coast of North America. Ph.D. dissertation, University of California, Davis.

Carlton, J. T. 1985. Transoceanic and interoceanic dispersal of coastal marine organisms: the biology of ballast water. Oceanography and Marine Biology: An Annual Review 23:313–371.

Carlton, J. T. 1999. The scale and ecological consequences of biological invasions in the world's oceans. Pp. 195–212 In: Invasive Species and Biodiversity Management. O. T. Sandlund, P. J. Schei, and Å. V. editors. Dordrecht: Kluwer Academic Publishers.

Carlton, J. T. 2001. Introduced Species in U.S. Coastal Waters: Environmental Impacts and Management Priorities. Arlington, VA: Pew Oceans Commission.

Carlton, J. T. 2007. Ballast. Pp. 249–251 In: The Oxford Encyclopedia of Maritime History, Volume 1. J. B. Hattendorf, editor. Oxford: Oxford University Press.

Carlton, J. T. 2009. Deep invasion ecology and the assembly of communities in historical time. Pp. 13–56 In: Biological Invasions in Marine Ecosystems. G. Rilov and J. A. Crooks, editors. Berlin: Springer-Verlag.

Carlton, J. T. 2011. Ballast. Pp. 43–49 In: Encyclopedia of Biological Invasions. D. Simberloff and M. Rejmanek, editors. Berkeley, CA: University of California Press.

Carlton J., and J. Geller. 1993. Ecological roulette: the global transport of nonindigenous marine organisms. Science 261:78–82.

Carlton, J. T., and G. Ruiz. 2005. Vector science and integrated vector management in bioinvasion ecology: conceptual frameworks. Pp. 36–58 In: Invasive Alien Species: A New Synthesis. H. A. Mooney, R. N. Mack, J. A. McNeely, L. E. Neville, P. J. Schei, and J. K. Waage, editors. Covelo, CA: Island Press.

Carlton, J. T., D. M. Reid, and H. van Leeuwen. 1995. Shipping Study. The role of shipping in the introduction of non-indigenous aquatic organisms to the coastal waters of the United States (other than the Great Lakes) and an analysis of control options. The National Sea Grant College Program/Connecticut Sea Grant Project R/ES-6. Department of Transportation, United States Coast Guard, Washington, DC, and Groton, CT. Report Number CG-D-11-95. Government Accession Number AD-A294809.

Chu, K. H., P. F. Tam, C. H. Fung, and Q. C. Chen. 1997. A biological survey of ballast water in container ships entering Hong Kong. Hydrobiologia 352:201–206.

Cohen, A. N., and J. T. Carlton. 1995. Biological Study. Nonindigenous Aquatic Species in a United States Estuary: A Case Study of the Biological Invasions of the San Francisco Bay and Delta. A Report for the United States Fish and Wildlife Service, Washington, DC, and The National Sea Grant College Program, Connecticut Sea Grant, NTIS Report Number PB96-166525.

David, M., S. Gollasch, M. Cabrini, M. Perkovic, D. Bosnjak, and D. Virgilio. 2007. Results from the first ballast water sampling study in the Mediterranean Sea – the Port of Koper study. Marine Pollution Bulletin 54:53–65.

D'Itri, F. M., editor. 1997. Zebra mussels and aquatic nuisance species. Boca Raton, FL: CRC Press.

Doblin, M. A., K. J. Coyne, J. M. Rinta-Kanto, S. W. Wilhelm, and F. C. Dobbs. 2007. Dynamics and short-term survival of cyanobacteria species in ballast water—implications for HAB invasions. Harmful Algae 6:519–530.

Drake, L. A. 2002. Ship-transported virio- and bacterio-plankton. Pp. 35–39 In: Commission Internationale pour l'Exploration Scientifique de la mer Mediterranee, Alien marine organisms introduced by ships in the Mediterranean and Black Seas, Istanbul, 6–9 November 2002, Workshop Monographs no. 20. Monaco.

Drake, L. A., K.-H. Choi, G. M. Ruiz, and F. C. Dobbs. 2001. Global redistribution of bacterioplankton and virioplankton communities. Biological Invasions 3:193–199.

Drake, L. A., A. E. Meyer, R. L. Forsberg, R. E. Baier, M. A. Doblin, S. Heinemann, W. P. Johnson, M. Koch, P. A. Rublee, and F. C. Dobbs. 2005. Potential invasion of microorganisms and pathogens via 'interior hull fouling': Biofilms inside ballast-water tanks. Biological Invasions 7:969–982.

Drake, L. A., M. A. Doblin, and F. C. Dobbs. 2007. Potential microbial bioinvasions via ships' ballast water, sediment, and biofilm. Marine Pollution Bulletin 55:333–341.

Duggan I., C. van Overdijk, S. Bailey, P. Jenkins, H. Limén and H. MacIsaac. 2005. Invertebrates associated with residual ballast water and sediments of cargo-carrying ships entering the Great Lakes. Canadian Journal of Fisheries and Aquatic Sciences 62:2463–2474.

Fahnenstiel, G., Y. Hong, D. Millie, M. Doblin, T. Johengen, and D. Reid. 2009. Marine dinoflagellate cysts in the ballast tank sediments of ships entering the Laurentian Great Lakes. Verhandlungen der Internationalen Vereinigung fur Theoretische und Angewandte Limnologie 30:1035–1038.

Ferguson, G. H. 1932. The chlorination of ballast water on Great Lake vessels. Public Health Reports 47:256–258.

Fofonoff, P. W., G. M. Ruiz, A. H. Hines, B. D. Steves, and J. T. Carlton. 2009. Four centuries of biological invasions in tidal waters of the Chesapeake Bay region. Pp. 479–506 In: Biological Invasions in Marine Ecosystems. G. Rilov and J. A. Crooks, editors. Berlin: Springer-Verlag.

Gollasch, S., E. MacDonald, S. Belson, H. Botnen, J. T. Christensen, J. P. Hamer, G. Houvenaghel, A. Jelmert, I. Lucas, D. Masson, T. McCollin, S. Olenin, A. Persson, I. Wallentinus, L. P. M. J. Wetsteyn, and T. Wittling. 2002. Life in ballast tanks. Pp. 217–231 In: Invasive aquatic species of Europe, distribution, impacts, and management. E. Leppakoski, S. Gollasch, and S. Olenin, editors. Dordrecht: Kluwer Academic Publishers.

Gray, D., and H. J. MacIsaac. 2010. Do zooplankton eggs remain viable despite exposure to open-ocean ballast water exchange: evidence from *in situ* exposure experiments. Canadian Journal of Fisheries and Aquatic Sciences 67:256–268.

Gray, D. K., T. Johengen, D. F. Reid, and H. J. MacIsaac. 2007. Efficacy of open-ocean ballast water exchange as a means of preventing invertebrate invasions between freshwater ports. Limnology and Oceanography 52:2386–2397.

Hallegraeff, G. M., and C. Bolch. 1992. Transport of diatom and dinoflagellate resting spores in ships' ballast water: implications for plankton biogeography and aquaculture. Journal of Plankton Research 14:1067–1084.

Hallegraeff, G. M. 1998. Transport of toxic dinoflagellates via ships' ballast water: bioeconomic risk assessment and efficacy of possible ballast water management strategies. Marine Ecology Progress Series 168:297–309.

Hines, A. H., G. M. Ruiz, and L. S. Godwin. 2000. Assessing the risk of non-indigenous species invasion in a high latitude ecosystem: ballast water treatment facility in Port Valdez, Alaska. Pp. 81–88 In: Marine Bioinvasions, Proceedings of the First National Conference, MIT Sea Grant College Program, Cambridge. J. Pederson, editor. Cambridge, MA: Massachusetts Institute of Technology.

Holeck, K., E. L. Mills, H. J. MacIsaac, M. Dochoda, R. I. Colautti and A. Ricciardi. 2004. Bridging troubled waters: understanding links between biological invasions, transoceanic shipping, and other entry vectors in the Laurentian Great Lakes. BioScience 10:919–929.

International Maritime Organization (IMO). 2004. International Convention for the Control and Management of Ship's Ballast Water and Sediments. International Maritime Organization, London, England.

Kaluza, P., A. Kolzsch, M. T. Gastner, and B. Blasius. 2010. The complex network of global cargo ship movements. Journal Royal Society Interface 7:1093–1103.

Kasyan, V. V. 2010. Holoplankton of ship ballast water in the Port of Vladivostok. Russian Journal of Marine Biology 36:167–175.

Kelly, D. W., G. A. Lamberti, and H. J. MacIsaac. 2009. The Laurentian Great Lakes as a case study of biological invasion. Pp. 205–225 In: Bioeconomics of Invasive Species. R. P. Keller, D. M. Lodge, M. A. Lewis, and J. F. Shogren, editors. Oxford: Oxford University Press.

Klein, G., K. MacIntosh, I. Kaczmarska, and J. M. Ehrman. 2010. Diatom survivorship in ballast water during trans-Pacific crossings. Biological Invasions 12:1031–1044.

Kirk, T. W. 1893. On the occurrence of *Xanthium strumarium*, Linn., in New Zealand. Transactions of the New Zealand Institute 26:310–313.

Lee, H., II, D. A. Reusser, M. Frazier, and G. Ruiz. 2010. Density matters: review of approaches to setting organism-based ballast water discharge standards. EPA/600/R-10/031. EPA Office of Research and Development, National Health and Environmental Effects Research Laboratory, Western Ecology Division.

Levings C., J. Cordell, S. Ong, and G. Piercey. 2004. The origin and identity of invertebrate organisms being transported to Canada's Pacific coast by ballast water. Canadian Journal of Fisheries and Aquatic Sciences 61:1–11.

Lavoie, D. M., L. D. Smith, and G. M. Ruiz. 1999. The potential for intracoastal transfer of non-indigenous species in the ballast water of ships. Estuarine, Coastal and Shelf Science 48:551–564.

Lodge, D. M., S. L. Williams, H. J. MacIsaac, K. R. Hayes, B. Leung, S. Reichard, R. N. Mack, P. B. Moyle, M. Smith, D. A. Andow, J. T. Carlton, and A. McMichael. 2006. Biological invasions: recommendations for U.S. policy and management. Ecological Applications 16:2035–2054.

McCarthy, H. P., and L. B. Crowder. 2000. An overlooked scale of global transport: phytoplankton species richness in ships' ballast water. Biological Invasions 2:321–322.

Miller, A. W., T. Huber, M. S. Minton, and G. M. Ruiz. 2010. Status and trends of ballast water management in the United States: Third Biennial Report to the United States Coast Guard. Edgewater, MD: National Ballast Information Clearinghouse, Smithsonian Environmental Research Center

Mills, E. L., J. H. Leach, J. T. Carlton, and C. L. Secor. 1993. Exotic species in the Great Lakes: A history of biotic crises and anthropogenic introductions. Journal of Great Lakes Research 19:1–54.

Minton, M. S., E. Verling, A. W. Miller, and G. M. Ruiz. 2005. Reducing propagule supply and coastal invasions via ships: effects of emerging strategies. Frontiers in Ecology and the Environment 3:304–308.

Radziejewska, T., P. Gruszka, and J. Rokicka-Praxmajer. 2006. A home away from home: a meiobenthic assemblage in a ship's ballast water tank sediment. Oceanologia 48:259–265.

Reid, D. F., T. H. Johengen, H. MacIsaac, F. Dobbs, M. Doblin, L. Drake, G. Ruiz, and P. Jenkins. 2007. Identifying, verifying, and establishing options for best management practices for NOBOB vessels. Report to the Great Lakes Protection Fund.

Ricciardi, A. 2006. Patterns of invasion of the Laurentian Great Lakes in relation to changes in vector activity. Diversity and Distributions 12:425–433.

Ruiz, G. M., P. W. Fofonoff, J. T. Carlton, M. J. Wonham, and A. H. Hines. 2000a. Invasion of coastal marine communities in North America: apparent patterns, processes, and biases. Annual Review of Ecology and Systematics 31:481–531.

Ruiz, G. M., T. K. Rawlings, F. C. Dobbs, L. A. Drake, T. Mullady, A. Huq and R. R. Colwell. 2000b. Global spread of microorganisms by ships. Nature 408:49–50.

Ruiz, G. M., and D. F. Reid, editors. 2007. Current state of understanding about the effectiveness of ballast water exchange (BWE) in reducing aquatic nonindigenous species (ANS) introductions to the Great Lakes Basin and Chesapeake Bay, USA: Synthesis and analysis of existing information. NOAA Technical Memorandum GLERL-142, National Ocean and Atmospheric Administration, Ann Arbor.

Santagata, S., and G. Ruiz. 2007. Task 3: Characterize source invertebrate populations and assess salinity toxicity as a barrier to prevent transfers of high-risk species to the Great Lakes in ballast tanks. Pp. 67–96 In: A final report for the project "Identifying, verifying, and establishing options for best management practices for NOBOB Vessels." Reid, D. F., T. Johengen, H. J. MacIsaac, F. C. Dobbs, M. Doblin, L. Drake, G. M. Ruiz, P. T. Jenkins, S. Santagata, C. van Overdijk, D. Gray, S. Ellis, Y. Hong, Y. Tang, F. Thomson, S. Heinemann, and S. A. Rondon (editors). Submitted to The Great Lakes Protection Fund by the University of Michigan, Cooperative Institute for Limnology and Ecosystems Research, and National Oceanic and Atmospheric Administration, Great Lakes Environmental Research Laboratory, Ann Arbor. Available online at: *http://www.glerl.noaa.gov/res/Task_rpts/2004/ais-reid04-1.html*. Last accessed June 2, *2011*.

Santagata, S., Z. R. Gasiunaite, E. Verling, J. R. Cordell, K. Eason, J. S. Cohen, K. Bacela, G. Quilez-Badia, T. H. Johengen, D. F. Reid, and G. M. Ruiz. 2008. Effect of osmotic shock as a management strategy to reduce transfers of non-indigenous species among low-salinity ports by ships. Aquatic Invasions 3:61–76.

Smith, L. D., M. Wonham, L. McCann, G. Ruiz, A. Hines, and J. Carlton. 1999. Invasion pressure to a ballast-flooded estuary and an assessment of inoculant survival. Biological Invasions 1:67–87.

Solow, A., and C. Costello. 2004. Estimating the rate of species introductions from the discovery record. Ecology 85:1822–1825.

Sun, B., R Mouland, C. Way, and R. B. Rivkin. 2010. Redistribution of heterotrophic prokaryotes through ballast water: A case study from the west coast of Canada. Aquatic Invasions 5:5–11.

Verling, E., G. M. Ruiz, L. D. Smith, B. Galil, A. W. Miller, and K. R. Murphy. 2005. Supply-side invasion ecology: characterizing propagule pressure in coastal ecosystems. Proceedings of the Royal Society B, 272:1249–1257.

Williams, R., F. Griffiths, E. van der Wal, and J. Kelly. 1988. Cargo vessel water as a vector for the transport of nonindigenous marine species. Estuarine and Coastal Shelf Science 26:409–420.

Wonham, M. J., J. T. Carlton, G. M. Ruiz, and L. D. Smith. 2000. Fish and ships: relating dispersal frequency to success in biological invasions. Marine Biology 136:1111–1121.

Wonham, M. J., W. C. Walton, G. M. Ruiz, A. M. Frese, and B. S. Galil. 2001. Going to the source: role of the invasion pathway in determining potential invaders. Marine Ecology Progress Series, 215:1–12.

Wonham, M. J., and J. T. Carlton. 2005. Trends in marine biological invasions at local and regional scales: the Northeast Pacific Ocean as a model system. Biological Invasions 7:369–392.

Wonham, M. J., M. A. Lewis, and H. J. MacIsaac. 2005. Minimizing invasion risk by reducing propagule pressure: a model for ballast-water exchange. Frontiers in Ecology and the Environment 3:473–478.

Zvyagintsev, A. Yu., V. V. Ivin, I. A. Kashin, T. Yu. Orlova, M. S. Selina, V. V. Kasyan, O. M. Korn, E. S. Kornienko, V. A. Kulikova, I. P. Bezverbnaya, L. V. Zvereva, V. I. Radashevsky, L. S. Belogurova, A. A. Begun, and A. N. Gorodkov. 2009. Acclimation and introduction of hydrobionts ships' ballast water organisms in the Port of Vladivostok. Russian Journal of Marine Biology 35:41–52.

2
Policy Context for Regulating Live Organisms in Ballast Discharge

The policy context around regulation of ballast water discharge is important to understanding the potential for a numeric standard on living organisms in ballast to reduce the risk of invasions. Two independent agencies, the U.S. Environmental Protection Agency (EPA) and the U.S. Coast Guard (USCG), have authority to directly regulate ballast water discharges to waters of the United States. How their programs developed, overlap, and complement one another is the subject of this chapter.

STATUTORY BACKGROUND OF BALLAST MANAGEMENT

The first United States law with regulatory jurisdiction over live organisms in ballast water discharge was, by a 2005 U.S. District court decision, the Clean Water Act Amendments of 1972 (Clean Water Act, 33 U.S.C. §§ 1251–1387, 2006). The Act and its subsequent amendments (hereafter referred to as the CWA) impose progressively more stringent requirements on point sources of pollutants via the National Pollution Discharge Elimination System (NPDES).

In 1973 when the EPA promulgated regulations implementing the NPDES, there was little or no attention from Congress or the environmental community to the specific problem of live organisms in discharges from vessels. In fact, EPA's regulations specifically exempted "discharge incidental to the normal operation of a vessel" (40 C.F.R. § 122.3(a) (2008)). This regulatory exemption, which greatly expanded a statutory exemption of "discharges incidental to normal operations of vessels *of the Armed Forces*" (emphasis added) (33 U.S.C. § 1362(6)), was considered unremarkable for over two decades. Environmental groups finally petitioned EPA to vacate the exemption in 1999. The agency denied the petition, and the groups filed suit. In March 2005, the U.S. District Court for the Northern District of California in San Francisco ruled that the EPA must begin regulating ballast water discharges through the CWA NPDES program (Northwest Envtl Advocates et al. *vs.* United States EPA, No. C 03–05760–SI, 2005). EPA issued its first Vessel General Permit (VGP) in 2008, and subsequent versions are anticipated that include water quality standards for ballast water.

Ballast introductions of aquatic invasive species did become salient in Congress in the late-1980s when massive infestations of the ship-mediated zebra mussel (*Dreissena polymorpha*) and Eurasian ruffe *(Gymnocephalus cernua*) burgeoned in the Great Lakes. Industrial and municipal raw water users and fisheries interests raised frantic appeals for federal action (Cangelosi, 2003). However, at that time, Congress and its constituents, including some in the environmental community, regarded most aquatic invasive species impacts to be economic rather than environmental. That is, even in the immediate wake of these dramatic infestations, there were few if any calls for the issue to be addressed under the CWA. Consequently, Congress sought in 1990 to fill an apparent regulatory vacuum with the Nonindigenous Aquatic Nuisance Prevention and Control Act of 1990, also known as NANPCA (P.L. 101-646; 16 U.S.C. 4701 et seq.,), which ordered the USCG to regulate ballast operations of ships.

NANPCA was the first federal law globally with the explicit purpose of regulating ballast discharges of aquatic invasive species. NANPCA required the USCG to issue voluntary guidelines within six months and regulations within two years of enactment to prevent new introductions of aquatic nonindigenous species by ships entering the Great Lakes. Modeled after a voluntary Canadian program, the new U.S. regulatory program required all ships entering the Great Lakes and northern sections of the Hudson River, after operating outside the Exclusive Economic Zone (EEZ) of the United States and Canada of 200 nautical miles from shore, to undertake ballast management operations prior to discharging ballast water in a U.S. Great Lakes port, or face criminal and civil penalties. Because all vessels entering the Great Lakes via the St. Lawrence River pass through U.S. territorial water, this regulation effectively covered all ballasted transoceanic vessels entering the Great Lakes system irrespective of whether their destination port was in the U.S. or Canada.

Specifically, NANPCA required ships to undertake open-ocean ballast water exchange, the only available approach at the time to reduce the risk of ship-mediated introductions of invasive species, or an alternative environmentally sound treatment method approved by the USCG as being at least as protective as ballast water exchange. NANPCA required that after a voluntary period of two years the program must become mandatory. It also set up an Aquatic Nuisance Species Task Force and regional panels for coordinating action to control aquatic invasive species infestations.

In 1996, Congress enacted the National Invasive Species Act of 1996 [NISA, Pub. L. No. 104-332, 110 Stat. 4073 (codified as amended 16 U.S.C.)] to reauthorize and expand the NANPCA program. The provisions of NISA reflect the extent to which the geographic scope of concern over ship-mediated introductions of invasive species had increased in just a half decade. NISA charged the USCG with promulgating a national ballast management program to complement the Great Lakes regional program. The law stipulated that the national program must be made mandatory within three years of its initiation if voluntary implementation was not adequate. To assure informed agency follow-through, Congress required vessels to report on ballast management activities carried out pursuant to the law from the time of enactment, and the USCG to periodically assess compliance based on these reports.

In every region except the Great Lakes, where ballast management was already mandatory, the maritime industry failed to report, much less implement, ballast management measures. In 2001, the USCG reported to Congress that the proportion of vessels that provided information on ballast water management was around 25 percent, too low for a positive assessment of compliance with the voluntary guidelines, making a transition to a regulatory program of ballast management nationally a foregone conclusion (USCG, 2001). The USCG fulfilled its statutory obligation to make the guidelines mandatory three years later, in July 2004.

Era of Ballast Water Exchange

Invasions by aquatic nonindigenous species discovered subsequent to ballast water exchange requirements on ships entering the Great Lakes, such as by the fishhook water flea *Cercopagis pengoi* in Lake Ontario in 1998, suggested that the shipping vector was still transmitting nonindigenous aquatic species into the Great Lakes. One major cause of continued problems was likely incomplete application of ballast water exchange. In the late 1980s, while policy was still being developed to address ship-mediated invasions, it was common for ships to enter the Great Lakes entirely loaded with ballast water, discharge the water in the upper lakes, and depart the lakes entirely loaded in cargo (grain). But by the mid-1990s, concurrent with, but not a consequence of, the advent of the first ballast water exchange regulatory program, a large percentage of vessels entered the St. Lawrence Seaway laden at least partially with cargo (e.g., steel) (Cangelosi and Mays, 2006), such that most vessels had less than 10 percent of their tanks loaded with declarable (i.e., pumpable) water (Colautti et al., 2003).

Ballast water exchange at that time was considered safe and practicable only on fully ballasted tanks. Thus, the USCG exempted ships that were carrying cargo rather than ballast from ballast water exchange. Ships had the option of declaring their condition as "No Ballast On Board" (NOBOB) and those that did were not subject to any further scrutiny. How little ballast constituted eligibility for the NOBOB exemption was not clearly defined, monitored, or documented. As a consequence, it is possible that some ships laden mostly with cargo but with a full unexchanged ballast tank or two declared NOBOB in good faith and proceeded to discharge their unexchanged ballast at ports of call in the Great Lakes in the 1990s and the first half of the following decade. A bi-national monitoring program was finally initiated by the U.S. and Canada to check 100 percent of ballast tanks for these vessels on the Great Lakes in 2006.

In addition to the problems attendant to unexchanged ballast water from this minority of full tanks discharged into the lakes, subsequent studies revealed that even the so-called "empty" tanks, carrying only unpumpable water, entrained a diverse but varied flora and fauna, and themselves could be a potentially important vector for new invasive species introductions to the Great Lakes. These ballast residuals were suspected as posing a risk owing to their predominance within the foreign trade, the live biota they contained, and the abundance of viable resting stages contained in residual sediments (e.g., Bailey et al., 2003; Reid et al., 2007).

The Shipping Federation of Canada, whose member companies transmit 95 percent of trade in Eastern Canada, agreed to (among other things) more rigorous ballast water management practices that subject residual ballast water to similar flushing practices as filled tanks, to purge and kill live freshwater organisms, and to minimize sediment accumulation. This was codified in the September 2000 Code of Best Practices for Ballast Water Management for vessels entering the Great Lakes. In 2002, the twin Seaway authorities—the St. Lawrence Seaway Development Corporation (USA) and the St. Lawrence Seaway Management Corporation (Canada)—required all foreign flagged vessels to comply with these best management practices. Transport Canada and the USCG later reinforced these regional requirements.

However, despite advancements in scope and accountability, the shipping industry generally complained that ballast water exchange was difficult to implement and costly in terms of fuel and carbon emissions; federal officials found ballast water exchange difficult to enforce; and scientists expressed concern that it was not adequate to abate continued invasions. Congress (P.L. 104-332), the International Maritime Organization (IMO), the USCG, and the National Research Council (NRC, 1996) began commending *ballast water treatment* in lieu of ballast water exchange as the best long-term solution to ballast-mediated aquatic species invasions. Concurrently, early treatment performance demonstrations suggested that mechanical operation of ballast treatment on ships could be feasible and biologically effective (Cangelosi, 1997; Parsons and Harkins, 2002; Rigby et al., 1997). With the acceptance of a future transition to routine ballast water treatment, interest in a performance standard that would be applicable to ballast water advanced, and a protracted exploration of options for setting measurable performance standards for ballast water management pursuant to NANPCA and NISA began.

STANDARD-SETTING PROCESSES OF THE TWO STATUTES

The authorities of EPA and USCG to regulate ballast water have not been formally coordinated, although Congress made an effort to do so in 1996 when it enacted NISA. It is therefore in the hands of the implementing agencies to effect that coordination for the benefit of the regulated community. To this end, EPA's Office of Wastewater Management and the USCG are working jointly to derive numeric limits for living organisms in ballast water for the next CWA Vessel General Permit and for regulatory programs of the USCG to prevent the establishment of new aquatic nonindigenous species through the discharge of ships' ballast water.

All actions pursuant to the Clean Water Act must support this statute's overall goal to "eliminate discharge of pollutants" into navigable waters; its interim goal to achieve, where attainable, "water quality which provides for the protection and propagation of fish, shellfish, and wildlife and provides for recreation in and on the water"; and its stated objective to "restore and maintain the chemical, physical, and biological integrity of the Nation's waters." Finally, the CWA contains a narrative standard, prohibiting the discharge of "toxic pollutants in toxic amounts." The CWA defines "pollutant" to include "biological materials" (33 U.S.C. 1362(6)) and toxic pollutants as "those pollutants, or combinations of pollutants, *including disease-causing agents*[4], which after discharge and upon exposure, ingestion, inhalation or assimilation into *any organism*, either directly from the environment or indirectly by ingestion through food chains, will, on the basis of information available to the Administrator, cause death, disease, behavioral abnormalities, cancer, genetic mutations, physiological malfunctions (including malfunctions in reproduction) or physical deformations, in such organisms or their offspring" (33 U.S.C. 1362(13)). Thus, these narrative standards also are applicable to organisms and human and wildlife disease agents that may be entrained in ballast discharge.

To make these goals and objectives operational, EPA and the states have developed water quality standards that address the pollutant limits needed to protect the natural resources of concern. These criteria reflect the latest scientific knowledge on "the kind and extent of all identifiable effects on health and welfare including, but not limited to, plankton, fish, shellfish, wildlife, plant life, shore lines, beaches, aesthetics, and recreation which may be expected from the presence of pollutants…"; "the concentration and dispersal of pollutants, … through biological, physical, and chemical processes"; and "the effects of pollutants on biological community diversity, productivity, and stability...for varying types of receiving waters" [33 U.S.C. 1314]. All of the above considerations are potentially relevant to biological constituents in ballast discharge.

NPDES permits are written to prevent point source discharges from violating water quality standards, using two basic types of effluent limits. In the case of ship discharges (and many other point sources), EPA has relied on technology-based standards based on the best available technology economically achievable (BAT) [33 U.S.C. 1311(a)]. Although technological, operational, and economic feasibility are considerations in the agency's decision-making, the guiding principle is that the BAT must result in "reasonable further progress toward the national goal of eliminating the discharge of all pollutants" [33 U.S.C.

[4] Case law supports an interpretation of "disease-causing agents" that includes bacteria and viruses. *See APWU v. Potter*, 343 F.3d 619 (2d Cir. 2003) (in a case involving CERCLA, which defines the term "pollutant or contaminant" to include "disease-causing agents," the court identified anthrax, a bacterium, as a "pollutant or contaminant"); *U.S. v. City of North Adams, Mass.*, 777 F.Supp. 61, 81 n.22 (D. Mass. 1991) (in a case involving the Safe Drinking Water Act, the court identified bacteria and fungi as "disease causing agents").

1311(b)(2)(A)]. The BAT should evolve with technological and commercial advancement of systems, but this reality is dependent upon agency follow-through.

When application of the EPA-designated BAT is not sufficient to meet water quality standards, the EPA must issue water quality-based effluent limitations, which consider site-specific evaluation of the discharge and its effect on a receiving system [see: *http://www.epa.gov/npdes/pubs/chapt_06.pdf*), 33 U.S.C. 1311(b)(1)(C)]. States may add conditions to NPDES permits if such conditions are necessary to assure that discharges will meet state water quality standards [33 U.S.C. 1341(d)]. Any ship that discharges in violation of an NPDES permit is subject to enforcement or suit by EPA, a state, or a citizen.

In comparison, NANPCA and NISA also speak to both environmental protection and feasible and available methods for achieving this goal, but the relationship is less structured than for the CWA. The narrative statutory goal of the Great Lakes-specific NANPCA program is "to prevent unintentional introduction and dispersal of nonindigenous species into waters of the United States through ballast water management and other requirements." The NANPCA definition of nonindigenous species, which subsumes both live organisms and pathogens, is "any species or other viable biological material that enters an ecosystem beyond its historic range, including any such organism transferred from one country into another." NANPCA authorized the Secretary of the Department of Transportation (acting through the USCG) to issue regulations specifically "to prevent the introduction and spread of aquatic nuisance species into the Great Lakes through the ballast water of vessels." "Aquatic nuisance species" are defined as "a nonindigenous species that threatens the diversity or abundance of native species or the ecological stability of infested waters, or commercial, agricultural, aquacultural, or recreational activities dependent on such waters." The voluntary national ballast management program created by NISA contained a similarly stated purpose: "to prevent the introduction and spread of nonindigenous species in waters of the United States by ballast water operations and other operations of vessels equipped with ballast water tanks."

In contrast to the CWA, NANPCA and NISA do not require development of discharge standards. They instead require that ships undertake specific ship operations, namely ballast water exchange consistent with USCG requirements or environmentally sound ballast water treatment that the USCG approves as being at least as effective as ballast water exchange, among others options. Thus, the statutes essentially set a "management standard" (ballast water exchange) as a performance floor. However, the USCG must infer a risk-reduction level associated with that management standard to determine an allowable ballast water treatment; the laws state an overall objective "to prevent introduction and spread" of aquatic invasive species and require the USCG to approve treatments based on their being "at least as effective" as exchange.

When the USCG began promulgating operational guidelines for "effective" ballast water exchange for the Great Lakes and national programs, it soon became clear that a metric of effectiveness based on some limit on numbers of near coastal organisms entrained in discharge, was all but impossible to construct and enforce. The ability of ballast water exchange to reduce numbers of potentially harmful organisms in ballast discharge is not a single value, but varies across ships, voyages, and even tanks within a ship. As discussed in Chapter 1, exchange dilutes and reduces initial densities of high-risk near-coastal organisms but not organisms generally, since open-ocean organisms (of lower risk) are taken up in the exchange as the near-coastal water is purged. The USCG ultimately imposed an indirect but measurable performance metric for ballast water exchange in its Great Lakes regulations of a necessary post-exchange ballast water salinity exceeding 30 parts per thousand (ppt), based on the assumption that highly saline water was indicative of effective open-ocean ballast water exchange. Unfortunately, this indirect approach is less reliable when applied to marine system voyages, due to high near-shore salinities at many harbors. Waite and Kazumi (2001) noted that a purge of 95 percent of the near-coastal water during ballast water exchange was close to optimal from a physical standpoint, offering an alternative indirect metric that would be applicable to all voyages. Methods for requiring and enforcing a 95 percent purge were incorporated into proposed reauthorization language for NISA as a national ballast water exchange performance metric and into the IMO D-2 standards (see later discussion).

For reasons of technical complexity, the USCG did not propose numeric metrics for allowable live organisms in exchanged ballast water for over two decades after the first law's enactment. However, such standards are now necessary in order to operationalize the NANPCA and NISA requirements that ships be allowed to substitute environmentally sound ballast water treatment "at least as effective as ballast water exchange" at preventing introduction and spread of aquatic invasive species by ships.

CURRENT INTERNATIONAL, FEDERAL, AND STATE STANDARDS

The USCG's first exploration of standards for ballast water treatment occurred in the context of discussions at the International Maritime Organization (IMO). Ballast water management came to the attention of the IMO as early as 1988, largely as a result of the Canadian government and the International Joint Commission/Great Lakes Fishery Commission report (IJC, 1990). In 1993, the IMO requested that all member nations implement voluntary guidelines based on the Canadian ballast management guidelines. In 1997, the Marine Environmental Protection Committee (MEPC) guidelines were reviewed, revised, and adopted as Assembly Resolution A 868(20). The resolution also requested the MEPC to work towards completion of a legally binding Convention, together with guidelines for uniform implementation.

For five years a U.S. interagency working group including both USCG and EPA participated in the MEPC's Ballast Water Working Group to develop the IMO's *International Convention for the Control and Management of Ships' Bal-*

last Water and Sediments. The Convention was adopted at the International Conference on Ballast Water Management for Ships in February 2004. Though the U.S. has not signed the agreement (Canada has), many convention provisions, including the standards, reflect critical input from both USCG and EPA during the deliberative process.

The Convention identifies two key standards, IMO D-1 and D-2, which vary with respect to minimum requirements, date of implementation, and application by size and date of build for different vessels (see Tables 2-1 and 2-2). D-1 is a ballast water exchange standard, requiring at least 95 percent volumetric exchange for empty-refill-style exchange, and three times the tank volume for vessels that conduct flow-through exchange. The D-2 standard sets maximum permissible limits on live organisms in ballast effluent, based on the size or taxonomic category of organisms (Table 2-2). It states that ships conducting ballast water management shall discharge:

- "Less than 10 viable organisms per m^3, for greater than or equal to 50 µm in minimum dimension;
- Less than 10 viable organisms per ml, for less than 50 µm in minimum dimension and greater than or equal to 10 µm in minimum dimension; and
- Discharge of the indicator microbes shall not exceed the specified concentrations. The indicator microbes, as a human health standard, include, but are not be limited to:
    - Toxicogenic *Vibrio cholerae* (O1 and O139) with less than 1 colony forming unit (CFU) per 100 ml or less than 1 CFU per 1 gram (wet weight) zooplankton samples;
    - *Escherichia coli* with less than 250 CFU per 100 ml;
    - Intestinal enterococci with less than 100 CFU per 100 ml."

The size classes contained in the IMO standard very roughly coincide with taxonomic groups of plankton (i.e., zooplankton and phytoplankton for the larger and smaller class sizes, respectively) in marine systems, but also with projected mechanical capacity to filter ballast water in a full-scale treatment process. The fecal indicator bacteria identified and the levels set for them reflected accepted human contact standards (EPA, 1986). The *V. cholerae* (serotypes O1 and O139) levels in the standard, a facet of the standard particularly hard to assess in treatment performance evaluations in the real world, were called out largely in response to concerns from Brazil that pathogens in ballast water pose a serious threat to receiving ports (MEPC 47/2/11, Brazil, 2003).

The United States delegation, along with some others, argued for a more stringent standard during IMO deliberations (BWM/CONF/14, 2004), but this position did not prevail. The U.S. delegation, backed by direct input from United States Senators (in a letter to Admiral Thomas H. Collins, dated January 22, 2004), did secure a provision in the convention that makes explicit an assumption that member states can unilaterally implement standards more stringent than

TABLE 2-1 IMO D-1 and D-2 Ballast Water Performance Standards

<table>
<tr><th>Standard</th><th colspan="2">Exchange Type</th><th colspan="2">Viable Organism Density</th></tr>
<tr><td rowspan="2">D-1</td><td>Empty-Refill</td><td>≥ 95% exchange</td><td colspan="2" rowspan="2">N.A.</td></tr>
<tr><td>Flow-through</td><td>3x tank volume</td></tr>
<tr><td rowspan="6">D-2</td><td colspan="2" rowspan="6">N.A.</td><td>Organism</td><td>Concentration</td></tr>
<tr><td>≥50 µm MD</td><td><10 m^{-3}</td></tr>
<tr><td>≥10 µm MD and <50 µm MD</td><td><10 ml^{-1}</td></tr>
<tr><td>Vibrio cholera</td><td><1 CFU 100 ml^{-1} or <1 g^{-1} zooplankton wet weight</td></tr>
<tr><td>Escherichia coli</td><td><250 CFU 100 ml^{-1}</td></tr>
<tr><td>Intestinal enterococci</td><td><100 CFU 100 ml^{-1}</td></tr>
</table>

N.A. = Not Applicable. MD = minimum dimension. CFU = colony forming unit.
Vibrio cholerae refers to toxic strains O1 and O139.

TABLE 2-2 Timeline for implementation of IMO Ballast Water Management Regulations

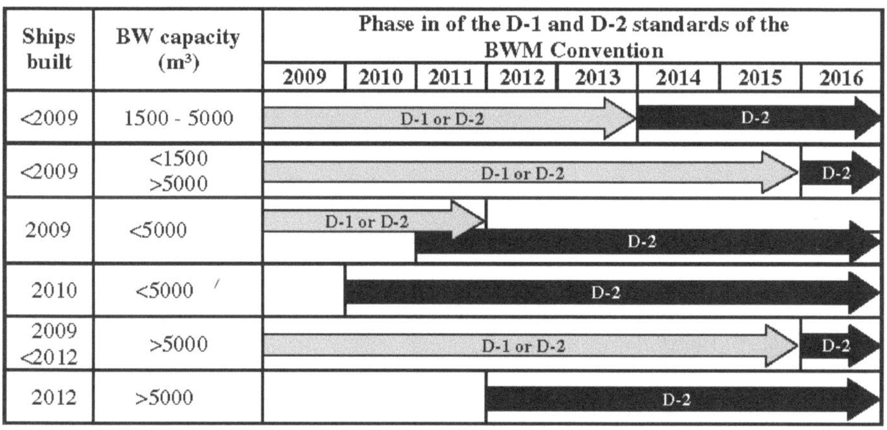

Note: Built refers to vessel build date. D1 refers to the need for vessels to conduct ballast water management that at least meets the ballast water exchange standards, while D2 refers to ballast water management that at least meets the ballast water performance standard. SOURCE: Reprinted, with permission, from David and Gollasch (2008). © 2008 by Stephan Gollasch.

IMO's. Canada ratified the Convention on April 9, 2010, while the United States, as yet, has not. The IMO Convention, which will come into force one year after not less than 30 States representing 35 percent of the world's merchant tonnage have ratified without reservation[5], nonetheless provides a common basis for the EPA and USCG to pursue domestic standards.

NISA-Related Requirements

In 2009, five years after the IMO Convention was agreed upon, the USCG issued a notice of proposed rulemaking citing its authority through NISA, which contained an interim and final numeric standard for live organisms in ballast discharge (Federal Register / Vol. 74, No. 166 / Friday, August 28, 2009). This action reflected demands derived in part from concerns that allowable levels of discharges of organisms within the IMO standard are not sufficiently different from taking no action. For example, the IMO discharge limit for protists of 10 live cells/ml does not significantly differ from even worse case uptake conditions. Indeed, an Australian study showed that a typical *Alexandrium* toxic plankton bloom may reach cell densities in ballast water of just one order of magnitude more than the IMO discharge limit, at 10^2 cells/ml (Hallegraeff, 2001). Others expressed concern that while the IMO standard will eliminate discharge of extremely dense concentrations of organisms, ballast discharge meeting the IMO discharge standard from a single ship could still represent a sizable inoculation when compounded over the numerous tons of ballast water a ship may load and subsequently discharge. Indeed, Bailey et al. (2009) postulated that the IMO D-2 standard (< 10 individuals m^{-3}) would result in a 0.27 probability of establishment for some zooplankton, like the parthenogenetic water flea *Daphnia retrocurva,* which because of its life history could colonize at very low inoculum densities; other zooplankton had lower or negligible invasion probabilities.

The proposed 2-phase USCG rule is shown in Table 2-3. The standard takes the potential for technological advancement into account by proposing an interim and final numeric standard which combine considerations of environmental protectiveness and technological feasibility. The proposed rule installs the IMO D-2 standard as an interim standard for ballast discharge, and a more stringent Phase 2 standard that ships must meet over time as technology becomes available. Phase 1 would come into effect immediately for *new* vessels constructed on or after January 1, 2012; for *existing* vessels constructed before January 1, 2012, the standard would come into effect in 2014 and 2016, depending on vessel ballast water capacity. Phase 2 would come into effect for all vessels on January 1, 2016, although existing vessels would have until their first

[5] Currently 28 countries representing over 28% tonnage have ratified the Convention.

TABLE 2-3 Phase 1 and Phase 2 Proposed USCG Ballast Discharge Standards

Phase	**> 50 µm**	**10 µm to 50 µm**	**Microbial (≤ 10 µm)**			
			Total	***V. cholerae****	***E. coli***	**enterococci**
1	< 10/ m^3	< 10/ml	NA	<1 CFU /100 ml	<250 CFU /100 ml	<100 CFU /100 ml
2	< 1/100m^3	< 1/100 ml	< 1,000 bacterial cells AND <10,000 viruses per 100 ml	<1 CFU /100 ml	<126 CFU /100 ml	<33 CFU /100 ml

*Toxicogenic *V. cholerae* O1 and O139

regularly scheduled dry dock after that date to install treatment technology. A final rule-making is pending. As discussed in Box 2-1, the virus standard (which is not part of the IMO D-2 standard) has been the subject of much controversy.

The USCG derived its proposed numeric limits for living organisms using the Population Viability Analysis approach (see Chapter 4). This model relates initial population size with extinction probability, which is known to be regulated by decreases in population size, low rates of population increase, and high variance in population size. The USCG concluded from this analysis (USCG DPEIS, 2008) that

- smaller inoculation density will yield less risk of establishment,
- treated discharge meeting the IMO standard will yield at least as much and usually more risk reduction than ballast water exchange, and
- risk reduction below that provided by the IMO D-2 standard is important enough to justify a second phase standard three orders of magnitude stricter than the IMO.

EPA Vessel General Permit

As noted above, the EPA has not yet developed numeric water quality criteria or identified best available technology economically achievable for live organisms in ballast water discharge.[6] The EPA's Vessel General Permit issued in

[6] While the federal component of the VGP contained no numeric limits on living organisms in ballast discharge, the state component of the process, the CWA section 401 certifications, resulted in several ballast water discharge standards added to the permit (see discussion below).

BOX 2-1
Viruses in the Context of Invasion Risk from Ballast Water

Aquatic viruses are very, very small; most range between 20 and 200 nanometers (1 nm = 10^{-9} meter). They also are very, very abundant; even in pristine coastal waters, their concentrations routinely range from 10^{10} to 10^{12} per liter. Suttle (2005) estimated the world ocean contains on the order of 10^{30} viruses. It is important to understand, however, that despite their enormous numbers, nearly every one of these viruses is of no consequence for human health. Overwhelmingly so, the host cells in which they propagate their genetic material are the myriad species of co-occurring bacterioplankton, phytoplankton, and heterotrophic protists found in fresh, brackish, and marine waters. In general, viruses are considered to be very host-specific, sometimes to the subspecies level (Lawrence, 2008). While there is potential for viruses pathogenic to humans to be released through ballasting operations, that possibility is less relevant than the potential introduction of viruses that infect fish, shellfish, and other ecologically or economically important aquatic species.

When dealing with environmental samples, there are a host of difficulties associated with sampling and concentrating viruses pathogenic to humans, then subsequently identifying them using cell-culture methods or reverse-transcriptase polymerase chair reaction (PCR). Enumeration of aquatic viruses is not straightforward. Even with transmission-electron microscopy (TEM), a technique that allows their direct visualization, there are problems attendant with recognizing some of their less distinctive morphotypes. A more commonly used method to enumerate aquatic viruses (more properly, "virus-like particles") involves staining their nucleic acids with a fluorochrome and counting pinpoints of light using epifluorescence microscopy (Patel et al., 2007). Neither staining nor TEM, however, can indicate whether the virus is infective or inactivated. Furthermore, visualization of viruses in seawater samples may overlook those already infecting host species and whose ballast water transport, therefore, would be tied to transport of the host bacterium, protist, fish, or shellfish.

As is the case with some free-living bacteria and protists, evidence is emerging consistent with the proposal that viruses have a biogeography. That is, not all viruses are distributed everywhere. Both metagenomic and observational studies have shown geographic constraints in the global-scale distribution of some viruses (studies cited in Lawrence et al., 2008).

Compared to other microorganisms, therefore, and certainly in contrast to metazoans, much has yet to be articulated about the biology and biogeography of aquatic viruses. There are a number of technical challenges that impede our improved understanding of them.

2008 contained only ballast water management and exchange requirements that largely mirrored the existing USCG regulatory program. However, a recent court settlement requires the agency to produce numeric criteria in draft by November 2011 in the next version of the VGP.

The current VGP includes some requirements not contained in the USCG program. For example, it disallows "discharges which violate water quality standards from vessels which are equipped with ballast tanks," referring to state and pending federal standards developed pursuant to the CWA. The VGP also specifies additional management requirements on vessels (unless the vessel meets one of the exemptions in Part 2.2.3.11 of the VGP)[7] engaged in Pacific nearshore voyages between Captain of the Port Zones, and on vessels with empty/unpumpable ballast residuals engaged in transoceanic or Pacific nearshore voyages. In these cases, various versions of salt-water flushing operations are required. Finally, the VGP stipulates that vessels may not discharge unexchanged or untreated ballast water or sediment into U.S. waters that are federally protected wholly or in part for conservation purposes.

Comparison of Regulatory Programs in EPA and USCG

Some features of the current EPA and USCG regulatory programs over live organisms in ballast discharge are inherently similar to each other. For example, both programs apply to ships equipped with ballast tanks, discharges of ballast water into "waters of the United States", and virtually any taxonomic category of organism in discharge, including viruses. Other features of the EPA and USCG existing programs are substantively different. The USCG program covers all aspects of the ship vector, including hull fouling and other sub-vectors, while the EPA program is focused on ballast discharges. The VGP applies to all vessels operating in a capacity as a means of transportation, except recreational vessels (as defined in CWA §502(25)) equal to or greater than 300 gross tons or with capacity to hold or discharge more than 8 cubic meters of ballast. This broad categorization of vessels includes some number of fishing vessels not included in the USCG program based on NISA. Alaska is not covered by the USCG, but it is covered by EPA.

Under the VGP, dischargers are responsible to conduct self-inspections and monitoring, and report on them annually, though the VGP does not designate a recipient for the reports. There is currently no inspection program associated with the VGP. Meanwhile, the USCG program includes mandatory reporting

---

[7] Exemptions (in Part 2.2.3.11 of the VGP) apply: Vessels may not conduct exchange/flushing for safety reasons, if an alternative, environmentally sound method of ballast water management approved by the USCG has been used prior to discharge, the vessel is in the Coast Guard STEP program, all ballast water is retained onboard (there is no discharge), the vessel does not traverse more than one COTP zone, and for Pacific nearshore voyages, the vessel is exempt if ballast water consists of water drawn from treated municipal water supplies, as long that ballast water is not mixed with any ballast water/sediments from other sources.

and a healthy system for receiving and archiving report information, and inspections to corroborate report claims. Statutory exemptions for the purposes of maintaining ship safety are similar for the two regulations, except that no exemptions are allowed for ships on the Great Lakes as required under NANPCA. The statutory features of agencies' standards are different in that EPA requires distinct water quality criteria and technology standards, and it sets, reviews, and revises its standards. The USCG technology standard is defined in the statute and requires risk reduction.

Finally, there are also relevant issues which neither the EPA or USCG ballast discharge regulatory program actively covers. One example involves oil tankers into Alaska, which are exempted by statute from requirements under NISA and also currently escape reporting to EPA because there is no inspection program associated with the VGP. (However, this latter exemption is not based on the statute and could be rectified with regulatory revision.)

State Standards for Ballast Discharges of Live Organisms

A number of U.S. states have implemented their own ballast water requirements via CWA Section 401 certification (see Table 2-4). These requirements emerged out of frustration with the perceived slow pace of federal developments; most states have indicated that they will relax or eliminate their programs should an effective federal program materialize.

On the Great Lakes, states with ballast water permitting policies include Michigan, Minnesota, New York, and Wisconsin. Michigan passed legislation in 2005 that requires all ocean-going vessels to obtain a permit from the Michigan Department of Environmental Quality before entering a port in state waters beginning January 2007. Permits for ballast water discharge are provided only for four state-approved ballast treatment technologies: sodium hypochlorite, chlorine dioxide, ultraviolet light, and deoxygenation. However, due to trade patterns, ocean-going vessels do not discharge ballast water in Michigan waters, only cargo. Consequently, no ship has been outfitted with treatment in response to this law.

In 2008, Minnesota passed legislation that required vessels to seek approval from the Minnesota Pollution Control Agency prior to discharging ballast water, and requiring existing vessels to treat to the IMO D-2 performance standards by 2016. Wisconsin began regulating discharges of ballast water by ocean-going vessels in February 2010, requiring 100 times the IMO D-2 performance standards for existing vessels by 2014. A subsequent technology availability review led the State to roll the requirement back to IMO D-2 performance standards provided they are undertaken in combination with ballast water exchange in the open ocean (see *http://dnr.wi.gov/news/mediakits/mk_ballast.asp*). Also in 2010, New York promulgated requirements that existing and new ships treat ballast to 100 and 1000 times, respectively, the IMO D-2 performance, even if they are simply passing through New York waters without discharging ballast water. New York recently issued a new deadline for compliance of 2013 (see *http://www.dec.ny.gov/docs/water_pdf/vesselgpletter.pdf*). Requirements on new ships are on a tighter time-line in all cases. Other Great Lakes states (e.g., Indiana, Illinois, Pennsylvania, Ohio) have less specific regulations, requiring a permit prior to ballast discharge by ocean-going vessels, and unspecified ballast management (Great Lakes Commission, 2008). Some Great Lakes' states include 'lakers' in their ballast water policies (Table 2-4).

The Rhode Island State Legislature charged the state's Department of Environmental Management with gathering information on international, federal, and state ballast water management policies and making recommendations for the establishment of a state ballast water management program. Other states such as Maryland and Virginia have laws requiring vessels to report their ballast operations. Oregon also has reporting requirements and requires vessels arriving from overseas to conduct ballast water exchange. California has a mandatory state ballast water management program that is supported by fees charged to vessel operators, and Washington State's program requires ballast water reporting and monitoring and establishes standards for ballast water treatment systems.

Other Countries

Transport Canada in the late 2000s amended the Canada Shipping Act to mandate ballast water exchange, treatment, and discharge to a reception facility, or retention on board for most foreign vessels arriving to Canada.[8] NOBOB vessels arriving from outside the Exclusive Economic Zone (> 200 nautical miles offshore in water $\geq$ 200m depth) are required to flush mid-ocean water through their tanks, treat ballast water to the IMO's D-2 performance standard, comply with the aforementioned best management practices, or demonstrate that proper ballast water exchange has previously been conducted to the IMO D-1 standard.

The Australian Quarantine and Inspection Service (AQIS) is designated as the lead agency for the management of ballast water risks in Australia. In 1990, AQIS introduced voluntary ballast water guidelines. The guidelines were refined and became mandatory on July 1, 2001. The requirements have legislative backing and are enforced under the Quarantine Act of 1908 (Commonwealth of Australia Quarantine Act 1908, as amended by Act No. 17 of 2002, *http://www.opbw.org/nat_imp/leg_reg/aus/qar.pdf*). There are no ballast water discharge standards distinct from the IMO standards associated with this Australian program.

[8] Excepting coastal vessels arriving from north of Cape Blanco (Oregon) or Cape Cod (Massachusetts), ballast water from these northern U.S. sources can be discharged only into receiving Canadian waters of the same biogeographic region (i.e., not into the Great Lakes).

TABLE 2-4 Summary of State Ballast Water Treatment Permit Requirements and U.S. Clean Water Act Certification Conditions as of January 2011

State	Regulatory Vehicle	Existing Oceangoing	New Oceangoing	Existing Lakers	New Lakers	Comments
California	California Code of Regulations Title 2, Division 3, Chapter 1	State discharge standard (more stringent than IMO) by 2014-2016, depending on vessel's ballast capacity	State discharge standard (more stringent than IMO) by 2010-2012, depending on vessel's ballast capacity			
Illinois	401 Certification	IMO by January 1, 2016	IMO for ships constructed on or after January 1, 2012	IMO by January 1, 2016	IMO for ships constructed on or after January 1, 2012	
Indiana	401 Certification	IMO by January 1, 2016	IMO for ships constructed on or after January 1, 2012	--	--	
Michigan	State permit; 401 Certification	Discharge prohibited without permit	Discharge prohibited without permit	--	--	Rights reserved to modify 401 Cert. if ballast treatment on Lakers is determined to be necessary, available and cost effective
Minnesota	State permit; 401 Certification	IMO by January 1, 2016	IMO for ships constructed on or after January 1, 2012	IMO by January 1, 2016	IMO for ships constructed on or after January 1, 2012	Requires MPCA approval of treatment technology

New York	401 Certification	100x by January 1, 2013	1000x IMO for ships constructed on or after January 1, 2013	100x IMO by January 1, 2012	1000x IMO for ships constructed on or after January 1, 2013	Vessel owners can request to extend compliance date if extension justified
Ohio	401 Certification	IMO by January 1, 2016	IMO for ships constructed on or after January 1, 2012	--	IMO for ships constructed on or after January 1, 2016	
Pennsylvania	401 Certification	IMO by January 1, 2016	Various standards more stringent than IMO for ships constructed on or after January 1, 2012	IMO by January 1, 2016	Various standards more stringent than IMO for ships constructed on or after January 1, 2012	Vessel owners can request to extend compliance date if extension justified
Wisconsin	State permit	IMO by January 1, 2014; prohibited from discharging sediment	IMO for ships constructed on or after January 1, 2012; prohibited from discharging sediment	Best Management Practices; prohibited from discharging sediment	Best Management Practices; prohibited from discharging sediment	

In New Zealand, voluntary controls were placed on the discharge of ballast water in March 1992 by the then Ministry of Agriculture and Fisheries. The controls sought to ensure that vessels refrain from discharging ballast water while in New Zealand waters. New Zealand introduced mandatory controls on ballast water discharges in 1998, under the Biosecurity Act 1993 (New Zealand Biosecurity Act 1993, Public Act 1993 No 95, *http://www.legislation.govt.nz/act/public/1993/0095/latest/DLM314623.html*).

CONCLUSIONS

The Clean Water Act and over 20 years of subsequent federal and state legislative activity explicitly addressing the problem of ship mediated introductions of invasive species have generated a complex domestic network of regulatory arrangements around ballast water discharge. Internationally, the regulatory arrangement is simpler, largely reflecting the IMO agreement terms, but implementation intensity is highly variable across member nations. This interagency, interjurisdictional regulatory mass is the starting point for deriving a numeric ballast discharge standard.

Fortunately, there are few relevant issues that the EPA and USCG regulatory programs, in combination, cannot actively cover. That is, though the current regulatory programs implementing the statutes have inadequacies, and future regulatory renditions may also fall short, **the statutes themselves appear to provide the essential considerations and scope to successfully implement scientifically based numeric standards, given sources of variation (described in Chapter 3).** For example, the CWA is capable of addressing place-specific issues, such as characteristics of the receiving system, through the State certification process and can even begin to address cumulative loading through the Total Maximum Daily Load (TMDL) program. Meanwhile, the NISA program is comprehensive with respect to the ship-related modes of introduction. Both statutes allow the implementing agency to be sensitive to critical risk factors such as voyage patterns and frequencies through variable enforcement intensity, and these considerations can be integrated into the EPA's water quality criteria development process. Both statutes can be complemented with other environmental laws to prevent problems with residual toxicity of treated discharge. The success of these programs will be contingent upon the willingness and ability of EPA and USCG to implement their regulatory programs to yield comprehensive coverage of risk factors in a manner that can be complied with by the regulated community.

Both EPA and USCG programs authorize rigorous enforcement of reporting and implementation actions by industry, a feature that will greatly facilitate data gathering on living organisms density and diversity that can support better understanding of the relationship between propagule pressure (in terms of inoculum density) and the probability of invasion (see Chapter 4). Specifically, the USCG regulations are likely to require ship owners to equip ships with sampling ports that will deliver representative samples of discharged ballast water (USCG Research and Development Center, 2008). Indeed, many of the few ships subject to treatment trials already have such sampling ports. The Great Ships Initiative is currently installing and trialing sample ports consistent with these guidelines in up to 12 Great Lakes-relevant ships. Once the sampling hardware is in place, and sampling protocols are agreed to, a major obstacle to generating high-quality and generalizable data for populating risk–release models will have been eliminated.

REFERENCES

Bailey, S. A., I. C. Duggan, C. D. A. van Overdijk, P. Jenkins and H. J. MacIsaac. 2003. Viability of invertebrate diapausing eggs collected from residual ballast sediment. Limnology and Oceanography 48:1701–1710.

Brazil. 2003. Ballast Water ANVISA. GGPAF Projects 2002. Pp. 3–4. Brazilia, February 2003. Available online at: *http://www.anvisa.gov.br/eng/pab/ballast_water3.pdf.* Last accessed on May 30, 2011.

BWM/CONF/14. 2004. Considerations of the Draft International Convention for the Control and Management of Ships' Ballast Water and Sediments: Ballast Water Discharge Standards, Regulation D-2.

Cangelosi, A., and N. Mays. 2006. Great Ships for the Great Lakes? Commercial Vessels Free of Invasive Species in the Great Lakes-St. Lawrence Seaway System. A Scoping Report for the Great Ships Initiative. Available online at: *http://www.nemw.org/scopingreport.pdf.* Last accessed June 1, 2011.

Cangelosi, A. 2003. Blocking Invasive Aquatic Species. Issues in Science and Technology, The National Academies, The University of Texas at Dallas, Winter 2002-03. Available online at: *http://issues.org*. Last accessed May 30, 2011.

Cangelosi, A. 1997. The Algonorth experiment. Seaway Review 25(3): January–March 1997.

Colautti, R. I., A. J. Niimi, C. D. A. van Overdijk, E. L. Mills, K. Holeck, and H. J. MacIsaac. 2003. Spatial and temporal analysis of transoceanic shipping vectors to the Great Lakes. Pp. 227–246 In: Invasive Species: Vectors and Management Strategies. G. M. Ruiz and J. T. Carlton, editors. Washington, DC: Island Press.

David, M., and S. Gollasch. 2008. EU shipping in the dawn of managing the ballast water issue. Marine Pollution Bulletin 56:1966–1972.

EPA. 1986. Bacteriological ambient water quality criteria for marine and fresh recreational waters. EPA 440/5-84-002. Cincinnati, OH: EPA Office of Research and Development.

Glassner-Shwayder, K., and T. Eder. 2010. Summary of Key Elements of Great Lakes State and Provincial Ballast Water Treatment Permit Requirements & U.S. Clean Water Act Sec. 401 Certification Conditions. Ann Arbor, MI: Great Lakes Commission.

Great Lakes Commission. 2008. Summary of Great Lakes State Ballast Water Legislation. Ann Arbor, MI: Great Lakes Commission.

Hallegraeff, G. M. 2001. Is a 99% Effective Ballast Water Treatment Sufficient? In: 1st International Ballast Water Treatment Standards Workshop Report. GloBallast Programme, 2001. Australia: University of Tasmania.

International Joint Commission. 1990. Exotic Species and the Shipping Industry: The Great Lakes/St. Lawrence Ecosystem at Risk: a special report to the governments of the United States and Canada by the International Joint Commission and the Great Lakes Fishery Commission.

Lawrence, J. E. 2008. Furtive foes: algal viruses as potential invaders. ICES J. Mar. Sci. 65:716–722.

Marine Environment Protection Committee (MEPC). 2001. 47/2/11. Harmful Aquatic Organisms in Ballast Water. Investigation carried out in selected ports in Brazil to identify and characterize pathogens in ballast water. Available online at: *https://www.ccaimo.mar.mil.br/sites/default/files/MEPC_47-2-11.pdf.* Last accessed June 2, 2011.

NRC (National Research Council). 1996. Stemming the Tide: Controlling Introductions of Nonindigenous Species by Ships' Ballast Water. Washington, DC: National Academy Press.

Parsons, M. G., and R. W. Harkins. 2000. The Great Lakes ballast technology demonstration project filtration mechanical test program. Marine Technology and SNAME News 37:129–140.

Parsons, M. G., and R. W. Harkins. 2002. Full-scale Particle Removal Performance of Three Types of Mechanical Separation Devices for the Primary Treatment of Ballast Water. Marine Technology 39(4):211–222.

Patel, A., R. T. Noble, J. A. Steele, M. S. Schwalbach, I. Hewson, and J. A. Fuhrman. 2007. Virus and prokaryote enumeration from planktonic aquatic environments by epifluorescence microscopy with SYBR Green I. Nature Protocols 2:269–276.

Reid, D. F., T. H. Johengen, H. MacIsaac, F. Dobbs, M. Doblin, L. Drake, G. Ruiz, and P. Jenkins. 2007. Identifying, verifying, and establishing options for best management practices for NOBOB vessels. Report to the Great Lakes Protection Fund. Availble online at: *http://www.glerl.noaa.gov/res/Task_rpts/2004/aisreid04-1.html.* Last accessed June 2, 2011.

Rigby, G., G. Hallegraeff, and C. Sutton. 1997. Ballast Water Heating and Sampling Trials on the BHP Ship MV Iron Whyalla in Port Kembla and en-route to Port Headland. Herausgeber AQIS. Archiv UTas eCite RODA Server (University of Tasmania Research Output Digital Asset Repository). Available online at: *http://ecite.utas.edu.au/10654.* Last accessed June 1, 2011.

Suttle, C.A. 2005. Viruses in the sea. Nature 437:356–361.

USCG (U.S. Coast Guard). 2001. Report to Congress on the Voluntary National Guidelines for Ballast Water Management, November 2001.

USCG. 2008. Ballast Water Discharge Standard. Draft Programmatic Environmental Impact Statement. April 2008.

USCG Research and Development Center. 2008. Analysis of Ballast Water Sampling Port Designs Using Computational Fluid Dynamics. CG-D01-08.

Waite, T., and J. Kazumi. 2001. Possible Ballast Water Treatment Standards: An Engineering Perspective. In: 1st International Ballast Water Treatment Standards Workshop Report. GloBallast Programme, 2001. Available online at: *http://globallast.imo.org/monograph%204%20standards%20workshop.pdf.* Last accessed June 2, 2011.

3
Sources of Variation Influencing the Probability of Invasion and Establishment

An implicit assumption in the desire of the U.S. Environmental Protection Agency (EPA), the United States Coast Guard (USCG), and other organizations to develop and refine a numeric standard for discharged ballast water is that there is a quantifiable relationship between the number of individuals of a given species released in a ballast discharge and the probability of its eventual establishment. These agencies have focused on inoculum density (ID, the number of organisms per unit volume of discharged ballast water), with the goal of setting the standard at a value(s) low enough to reduce or prevent the introduction of nonindigenous species. While a relationship between inoculum density and establishment probability may exist, many other factors beyond the simple density of released organisms affect establishment success in aquatic (freshwater and marine) systems. These factors include the identity (taxonomic composition), sources, and history of the propagules, and their abundance (total number of organisms), quality, and frequency of delivery. Further influencing the outcome of propagule release is a host of factors that include both species traits and the recipient region's environmental traits.

The following equation reflects the complex and uncertain relationship between propagule pressure, other variables, and the probability of establishment:

$$P_E = f(PP, \varepsilon)$$

where

P_E = probability of establishment of a self-sustaining population of a species

f = function

PP = propagule pressure over a given temporal–spatial scale

ε = a modifier that accounts for the many variables that can influence the survival, reproduction, and establishment of a species.

The factors encompassed by ε (sources of variation) are discussed here to underscore the importance of both event-specific (the moment of the release of a given inoculum from a single ship) and site-specific conditions when discussing invasion risk.

Figure 3-1 summarizes these sources of variation. The figure commences at the release or discharge of ballast water. At this stage, critical variables include (1) *inoculum abundance, density, and frequency,* (2) *the identity, diversity, source and history of the inoculating propagules*, and (3) *propagule quality*. Post-discharge processes then strongly influence the fate of the released propagules. These variables include both *species traits* and *environmental traits,* covering a very broad range of biological and ecological phenomena. These sources of variation shown in Figure 3-1 are reviewed below. This overview is not a comprehensive list of all factors that can influence invasion success, but is intended to illustrate significant sources of variation that are likely to influence the relationship between propagule supply and invasion outcome.

PROPAGULE PRESSURE

Inoculum Abundance, Density, and Frequency

A long-standing presumption in invasion biology is that an increase in inoculum abundance, density, and frequency will *increase* the probability of a species establishing a population in a new region. Thus, salmonid fish stocked at higher abundances and/or frequency are far more likely to colonize a new lake or reservoir than those that are not (Colautti, 2005). During World War II, the Korean War, and the Vietnam War, increased invasions occurred around the world correlated with increased shipping (and thus presumptive increase in species transportation and release) of those eras (Carlton and Norse, 2003). Numerous other such dose-response relationships are documented in the island biogeography, fisheries, and population biology literature (Cowx, 1997; Ricklefs and Miller, 1999; Losos and Ricklefs, 2009). Conversely, when inoculum abundance, density, and frequency *decrease*, a presumption is that the probability of species becoming established will also decrease. At its extreme, this is demonstrated by the history of vectors that cease to operate or are significantly reduced. During the Atlantic commercial oyster industry that brought oysters (*Crassostrea virginica*) from the United States east coast to the west coast (an episode that flourished between the 1870s and 1920s), numerous Western Atlantic bay and estuarine species were introduced to the North Eastern Pacific Ocean (Cohen and Carlton, 1995). When the industry ceased, documented invasions by that vector ceased. When Japanese oysters (*Crassostrea gigas*) were introduced to France in the 1970s, numerous western Pacific species colonized Europe (Goulletquer et al., 2002); when the industry ceased, documented invasions by that vector ceased.

In terms of the focus of this report, inoculum density in ballast water can vary hugely (Chapter 1). ID is usually reported as the number of organisms

STAGE IN THE INVASION PROCESS	SOURCES OF VARIATION OVER SPACE AND TIME
DISCHARGE (Release / Inoculation)	***Propagule Pressure:*** • **Identity, Diversity, Source and History of Propagules** • **Inoculum Abundance, Density, and Frequency** *Abundance*: N propagules *ID:* N propagules/ballast water volume) *Frequency:* ID / vessels / time • **Propagule Quality:** Viability and/or Reproductive Capability
POST DISCHARGE PROCESSES (Additional Variables that Regulate and Influence Survival, Reproduction, and Establishment)	***Species Traits:*** (in addition to **Propagule Quantity** and **Quality**, above) • **Genetics**: Bottlenecks, Quality, Variability, Enhancement, and Novelty • **Life History Characteristics** • **Population Density and Abundance** • **Habitat, Trophic, and Physiological Breadth** • **Dispersal and Mobility** • **Environmental Matching** (compatibility) ***Environmental Traits:*** ABIOTIC CHARACTERISTICS • **Habitat Landscape:** Scale of habitat fragmentation or connectivity • **Retention** of Propagules in Habitable Environment BIOTIC CHARACTERISTICS • **Biological Interactions** with Resident Species (native and nonindigenous) BIOTIC/ABIOTIC CHARCTERISTICS • **Disturbance Regimes**

FIGURE 3-1 Examples of variables altering establishment probability (P_E) of nonindigenous species released in ballast water and sediments. Processes in the right hand column impact the outcome of survival, reproduction, and establishment.

(propagules) per cubic meter of water, representing the number of individuals that could be released upon discharge. Ballast systems can also have attached fouling organisms on tank walls that form more-or-less permanent (in ballast tanks) and transient (in flooded cargo holds) communities, as well as benthic assemblages. In the case of these fouling and benthic organisms, their densities would be reported as per square meter. Fouling and benthic communities could contribute propagules (larvae) to the water, such that ballast systems would then essentially be self-seeded. Fouling and benthic organisms, often not quantified or characterized in ballast communities, could conceivably be released as well during deballasting.

As with inoculum density, inoculum frequency—the rate of propagule delivery per a given cohort of vessels over a given time period—can vary widely (see Chapter 1). Thus, combining the concepts of ID with propagule source and frequency, an enormous range of scenarios can potentially strongly influence invasion outcomes. Patterns can range from one inoculum of ballast water being released from one source at one time into one location to multiple inoculations from multiple sources released at multiple times in multiple locations. If environmental conditions are suitable, a single large, albeit episodic, inoculum (a so-called "pulse" event) may result in a successful invasion, even when many smaller inocula ("press" events), even if constant and steady, may not (see Bender et al., 1984). As noted in Chapter 4, small releases spaced sufficiently close in time can effectively resemble a single large release, in the form of a "cumulative load." Also noted in Chapter 4, and applicable here as well, is a "rescue" effect where multiple inocula act to sustain a nonindigenous population that might otherwise not have become established.

Source and History of Propagules and Their Effect on Species Identity, Diversity, and Abundance

Significant variation can exist in the characteristics of biota transported in ballast from different source regions. Even when controlling for total concentration (as, for example, with a discharge standard), species composition of a cubic meter of ballast water loaded from different sources can exhibit strong spatial and temporal variation. Source regions can be separated into distinct biogeographic zones (Ekman, 1953; Briggs, 1974; Udvardy, 1975) that are thought to experience differences in the capacity of their species to colonize a particular target location, assuming equal opportunity and similar habitat types (Mooney et al., 2005; Sax et al., 2005; Lockwood et al., 2007; Davis, 2009), such as the intertidal mudflat of a bay. The underlying mechanisms may include differences in physiological tolerance, competitive ability, and life history. For example, it has been suggested that biota from some geographic regions may be more successful in invading, compared to others, due to competitive superiority or particular suites of life history characteristics that have evolved in those regions (Vermeij, 1991, 1996). Numerous studies demonstrate asymmetries in the importance of different donor regions as sources of invasions (e.g., Carlton, 1992; Galil and Zenetos, 2002), but the extent to which this is driven by specific traits versus differences in opportunity (transport) or biotic and abiotic characteristics of recipient regions remains largely unexplored.

With respect to propagule history, there is considerable variation among ballast tanks and voyages in initial conditions (environmental and biotic), voyage durations/routes, and tank residence times that can affect the resident organisms. Past studies have shown that while some species can increase in abundance over time during the voyage (Carlton, 1985; Gollasch et al., 2000; Seiden et al., 2010), mortality rates (which can vary greatly among voyages) generally increase with residence time (Wonham et al., 2000; Verling et al., 2005).

Propagule Quality

Organisms are exposed to a variety of conditions during the transport process that can affect propagule quality. Propagules that arrive that are severely physiologically compromised (damaged, starved, or otherwise stressed) are not as likely to survive as those that have enjoyed luxurious transport conditions. For example, ballasted communities are often closed systems that change over time, such that the organisms present experience changes in temperature, water chemistry, and food that affect both physiological condition and survivorship. Of those organisms that survive transit to a recipient port, physiological condition upon arrival can affect the probability of establishment, in terms of both initial survival upon discharge and their eventual capacity to reproduce. Thus, the relationship between propagule supply and probability of establishment may differ greatly among individual discharge events, even when controlling for inoculation density. While there are likely strong effects of season, duration, source and many other factors that affect the condition of organisms upon arrival, with likely consequences for establishment probabilities, few data are available that quantify the condition of organisms upon arrival or resulting variation in performance.

SPECIES TRAITS

Species traits that influence the density and quality of propagules upon discharge (discussed above) include a host of characteristics that describe the life history, genetics, biology, and eventual population abundance of a species should it survive initial release. The likelihood of colonization can differ dramatically among species, reflecting a combination of both species-specific idiosyncratic characteristics as well as broader traits shared by related phylogenetic groups. It is easy to imagine how particular traits may confer advantages under certain circumstances, and there is a rich literature on this topic (Williamson,

1996; Ehrlich, 1986; Sax et al., 2005, Lockwood et al., 2006; Cadotte et al., 2006). While there are numerous examples of "matches" of life history traits, trophic modes, or other characteristics with the new environment in which species have been introduced, there is still a great deal of uncertainty—and thus lack of predictability—in aquatic ecosystems about which particular traits confer greater advantage. Nonetheless, one should expect difference among species in probability of establishment (controlling for conditions and propagule supply) due to variation in a wide range of biological and ecological traits.

Genetics: Bottlenecks, Quality, Variability, Enhancement, and Novelty

Invasion success may be related to a complex array of genetic phenomena. A released inoculum may have only a small proportion of the total genetic makeup (called founder effect) of the species in question—a bottleneck which may seriously inhibit the ability of the population to adapt to the range of environmental conditions found in a new location. Potentially overcoming this lowered genetic diversity is that multiple inocula of the species may be continuously released into the same location, either from the same source or from multiple sources; both could work to enhance genetic heterozygosity. The genetic quality of founding individuals (the "breeding value" of individuals for total fitness; Hunt et al., 2004) can be altered by transportation conditions. Novel strains of already-present species could also be introduced. In North America, the distribution and abundance of the native common reed grass *Phragmites australis* has increased dramatically since the 1800s. Genetic analysis revealed that an introduced genetic strain of *Phragmites* was responsible for the observed spread (Saltonstall, 2002; Vasquez et al., 2005). When the European green crab *Carcinus maenas*, introduced to the United States east coast in the early 1800s, began spreading north into Canada in the 1980s, it was assumed that *Carcinus* was simply expanding its range; instead, new genetic lineages from northern Europe had been separately introduced and were responsible for the northern appearance of this species (Roman, 2006). In a similar fashion, species long thought to be incapable of being introduced could become established because of the introduction of novel strains from new regions, due to altered trade routes.

Life History Characteristics

Invasion success can be profoundly influenced by life history characteristics. These include whether a species broods its young (direct development) versus releasing planktonic larvae, sexual versus asexual reproduction, and the ability to form resting stages. An example of direct development facilitating an invasion is the recent establishment of the North Atlantic rocky shore snail *Littorina saxatilis* in San Francisco Bay (Carlton and Cohen, 1998). This small periwinkle broods its young, which means that "miniature adults" can be released directly from a parent onto the rock substrate, with potentially high juvenile retention; the species can thus create dense populations of individuals that can easily locate and reproduce with each other. In contrast, newly released nonindigenous species that reproduce sexually and produce planktonic larvae are likely to have their larvae so widely dispersed that the likelihood of individuals surviving and settling out of the water adjacent to each other is low, unless inoculation and subsequent reproduction occur in a highly retentive environment, as discussed below.

Population Density and Abundance: Initial Post-Discharge Survival and the F_1 population

Propagules that have survived release may settle and grow in the new environment, and, if so, establish initial populations. At this point, this initial founder population—although composed of individuals that have grown to adulthood—faces extinction risk from many biological, ecological, physical, and chemical sources, prior to any of the individuals successfully reproducing. The founder population density is not the same as the inoculum density, which is the number of individuals *released* into the environment. Many of the individuals in the ID will perish upon release, leaving a proportion (which will vary across a vast spectrum of species and environmental regimes) to potentially survive, feed, and grow.

If the founding population survives initial extinction and produces a first generation (F_1), these new individuals, the first to be born in the new location, could, in turn, form dispersed metapopulations with a lowered probability of the newly colonizing species going extinct. The processes that influence the fates of both the initial founder population and F_1 reflect the entire suite of species traits and environmental traits already discussed that influence the probability of colonization, including the variables of demographic stochasticity discussed in Chapter 4.

Habitat, Trophic, and Physiological Breadth

A broad combination of ecological traits can result in a complex repertoire of advantages or disadvantages in colonization success (Ehrlich, 1986; Sax et al., 2005; Cadotte et al., 2006; Lockwood et al., 2006; Davis, 2009). Species that are eurytopic (broad habitat and physiological breadth) and euryphagic (broad feeding range) may have a higher likelihood of initial survival than species that are stenotopic and stenophagic (of narrow habitat, physiological, and feeding range). Alternatively, given the almost inconceivable number of circumstances that could greet released organisms, it is possible that particular functional groups, such as a certain trophic mode or a sedentary life-style, may be beneficial. Habitat, trophic, and physiological breadth of a colonizing species are closely linked to environmental compatibility and matching, discussed below. For example, species with more exacting food requirements may do well if their food is present, particularly those nonindigenous herbivores or carnivores that are introduced with, or at some point follow, their introduced plant and animal prey.

Dispersal and Mobility

A broad range of post-settlement dispersal mechanisms exist across thousands of species of protists, animals, and plants, ranging from passive means (such as entrainment on floating plants or anthropogenic debris—Thiel and Haye, 2006) to active means of locomotion (Levinton, 2008). This broad range of dispersal and mobility characteristics may influence initial colonization scenarios and the differential importance of metapopulation extinction. Thus, species that disperse to new locations relatively quickly and form new populations may have a higher probability of permanent establishment (and conversely a lower risk of extinction) than those species with low dispersal capability. Closely linked to the importance of dispersal and mobility is the role of propagule retention in a habitable environment, discussed below.

Environmental Matching

Upon delivery to a recipient region, a suite of local environmental factors may affect the probability of establishment, such that delivery of identical inocula to multiple locations or at different times can result in very different establishment outcomes. In other words, there is variation among sites in resistance or susceptibility to invasion (see Lonsdale, 1999, for detailed discussion of the components of invasion resistance). At its most extreme, the probability of establishment can be low or zero, even when large numbers of propagules are released, when the environment is inhospitable (for example due to severe mismatches in temperature or salinity).

There are often strong differences in species assemblages among different source habitats. The differences are perhaps most striking when comparing biota from open-ocean, coastal marine bays and estuaries, and freshwater, which can differ in taxonomic groups, functional groups, species life histories, and species environmental tolerances. Such large inter-habitat differences are thought to result in major barriers to invasions. Thus, the probability for oceanic or freshwater species to become established in the variable salinity of coastal bays and estuaries is generally considered low. Indeed, this is the rationale for the use of mid-ocean ballast water exchange (Ruiz and Reid, 2007, Santagata et al., 2008). As with all biota, there are exceptions, where some species can occupy a wide environmental range that extends across these disparate habitats.

It is evident that each species has limits to its geographic distribution imposed by environmental requirements and tolerances, defining its potential range; outside of this envelope, colonization cannot occur. While the specific environmental tolerances are not known for most marine and freshwater species, the likelihood of a successful invasion is expected to increase as environmental conditions approach those in its current range (i.e., when there is strong environmental matching). Thus, organisms transferred among biogeographic zones with similar environmental conditions (e.g., temperature regime) are more likely to find suitable conditions for colonization than those from dissimilar regions, controlling for other factors. On average, a decrease in the percentage of species that can colonize is expected as environmental similarity decreases.

A mismatch in environmental conditions can range from permanent to transient (controlling for the same inoculum). Transient mismatches can result from seasonal climatic variations, such as might occur for ballast being released in the winter versus summer, or daily weather variations, such as storms or unusual temperature conditions. More broadly, the environment can change over years, decades, or centuries, such that invasions previously excluded, despite long-term inoculation, can be successful at a later time. These changes encompass directional climatic changes (such as warming temperatures), other long-term environmental shifts that may or may not be related to climate change (such as changes in regional hydrography that would alter propagule retention), or macrocyclic events (such as El Niño – Southern Oscillations or decadal oscillations).

While there is general consensus that environmental matching is a key variable in invasion outcome, and many different models exist to predict the potential range of species (based on environmental matching) (Sax et al., 2005; Cadotte et al., 2006; Lockwood et al., 2006; Phillips and Dudik, 2008; Elith and Leathwick, 2009), many of the model outputs have not been rigorously tested, such that a precise understanding of variation in invasion probability as a function of environmental dissimilarity is not yet available.

ENVIRONMENTAL TRAITS

A number of abiotic and biotic processes play a significant role in determining the invasion success of organisms in discharged ballast water. These are discussed below, focusing first on abiotic characteristics of the recipient region, followed by biotic interactions between invaders and species in recipient regions, and finally the role of disturbance regimes.

Scale of Habitat Fragmentation or Connectivity

Several characteristics of coastlines and bays/estuaries are likely to influence the likelihood of establishment and spread of propagules introduced by ballast water discharge. These factors include the relative distance among bays/estuaries and appropriate recipient habitats along the outer coast, which can influence how connected these habitats are to each other. Increased continuity of invasible habitat, including artificial substrate (Glasby et al., 2007; Ruiz et al., 2009), will increase the connectivity of populations to the degree that propagules (e.g., larvae) can disperse with currents among other bays and invasible habitats. Increased connectivity can result in larger effective population sizes for invaders, and thus reduced likelihood of demographic stochasticity, inbreeding, or other problems associated with small populations. Conversely, highly fragmented habitats can lead to population isolation and smaller population sizes, with increased demographic stochasticity and increased inbreeding. The diversity of the habitats available with bays/estuaries will also be likely to increase the likelihood of establishment, due to increased likelihood of the presence of habitats with minimum requirements of food or shelter from predators.

An example of the importance of habitat continuity and connectivity on risk of establishment is presented by the European green or shore crab (*Carcinus maenas*) and the Chinese mitten crab (*Eriocheir sinensis*). The European green crab successfully colonized nearly 1,500 km of shoreline along the North American Pacific coast from Vancouver Island, British Columbia, to Monterey Bay, California, over an approximately ten year period after it was first discovered in San Francisco Bay around 1990 (Behrens Yamada et al., 2005). In contrast, since its discovery in the early 1990s, also in San Francisco Bay, the Chinese mitten crab has not expanded beyond the boundaries of the bay despite also having a planktonic dispersal stage. Unlike the European green crab, which is restricted to marine waters, mitten crabs require both marine and fresh water to complete their life cycle, returning to marine waters only for reproduction (Rudnick et al., 2005). The differences in life history coupled with the spatial separation of low salinity estuaries along the west coast may have provided an effective obstacle to subsequent spread for mitten crabs but not for green crabs.

Retention of Propagules in Habitable Environment: Diffusion vs. Retention

The probability of invasive species establishment and spread can be influenced by aspects of the local and regional oceanography and geomorphology of the recipient habitat, typically a bay or estuary (Crisp, 1958; Cohen et al., 1995). Among the many factors that are likely to influence this probability are residence time within the bay/estuary, magnitude and frequency of inputs from adjacent nearshore oceans, and inflow from local watersheds. These inputs will frequently dictate several physical properties of the bay/estuary that will influence the likelihood that propagules introduced through ballast water will become established. One is the residence time of a given parcel of ballast water, which will strongly influence the degree to which planktonic stages (larvae, eggs, small juveniles) are retained in the bay/estuary vs. advected out to offshore waters. A second set of factors influenced by ocean and watershed inputs is water column properties such as temperature, salinity, turbidity, nutrient levels, and pelagic food web components including phytoplankton and zooplankton. Overall, the greater the residency time of a parcel of water and the greater the hydrographic restrictions that would lead to decreased larval export from a suitable environment (Crisp, 1958), the greater the probability of the development and survival of initial and continuing populations.

Biological Interactions with Resident Species

The risk of establishment and subsequent spread can be influenced by a host of interactions between the invading propagules and organisms in the recipient habitat both native and nonindigenous. *Decreasing* the probability of establishment, for example, would be the presence of predators, competitors, pathogens, and parasites that might reduce the chances that newly introduced propagules from ballast discharge will result in an established population (Sakai et al., 2001). This Biotic Resistance Hypothesis (BRH), where invaders have no co-evolved defenses and competitive advantages and native consumers limit invasion, dates back to Elton (1958). The absence of species that would provide trophic support (for example, having no co-evolved prey capture strategies) or the absence of habitat created by other species (that might provide refugia) could also decrease the probability of establishment of ballast water invaders.

Increasing the probability of establishment could be the absence of predators and parasites that would normally curb an organism's growth and survival in its native range. Increased success that is the result of "release from enemies" is referred to as the Enemy Release Hypothesis (ERH) (Torchin et al., 2003). The probability of establishment is also aided by the naïve prey hypothesis (NPH), which argues that naïve prey may be less defended against new invasive predators (Parker et al., 2006; Sorte et al., 2010). A related idea suggests that nonindigenous species may possess "novel weapons" (e.g., phytochemicals in plants, novel modes of consumption) that native species cannot overcome (Callaway and Rindenour, 2004). Similarly, the absence of co-evolved competitors (the Competitive Release Hypothesis, CRH, formerly called the Evolution of Increased Competitive Ability, Blossey and Notzhold, 1995) could facilitate invasion, particularly if the species involved represent novel trophic modes in the recipient system. Further facilitating establishment could be the presence of species that would provide food or refugia for invading propagules. Finally, it has also been shown that the success of invaders may be facilitated by other invasions, for example, where an introduced consumer selectively reduces native prey, thus releasing an introduced competitor (Grosholz, 2005; Parker et al., 2006). This ability of one invader facilitating others could potentially result in a process referred to as invasional meltdown (Simberloff, 2006).

The relative importance of these hypotheses is debated, and evidence for them in marine systems is often equivocal (Parker et al., 2006). An interesting example of a food web interaction that has affected invasion success involves the salt marsh cordgrass *Spartina alterniflora*, native to the Western North Atlantic Ocean, and hybrids. The expansion of hybrid *Spartina* in San Francisco Bay and *Spartina alterniflora* in Willapa Bay, Washington, has resulted in dramatic shift towards a food web dominated by deposit feeders, many of them nonindigenous, at the expense of native surface-feeding worms and clams (Neira et al., 2005, 2007; Levin et al., 2006; Grosholz et al., 2009). Other invasive species in San Francisco Bay such as the North Atlantic mussel *Geukensia demissa* are strongly dependent in certain localities on the habitat created by *Spartina*. This facilitation appears strongest for invading species that have evolved in habitats where *Spartina* was present historically (Grosholz, unpublished data). This is a good example of how a species can modify marine and estuarine habitats and dramatically change the food web, in some cases increasing abundances of other nonindigenous species at the expense of native species.

Disturbance Regimes

Another factor that may influence the probability of establishment of propagules from ballast water is the issue of human disturbance (habitat modification, water quality changes, and so forth) and how this disturbance could influence the likelihood of both establishment and spread. Although this idea has a long history in invasion biology (Elton, 1958; Mack et al., 2000), studies of invasions in other systems have provided contradictory evidence that disturbance increases the likelihood of establishment (Simberloff, 1989; Case, 1996; Stohlgren et al., 1999; Mack et al., 2000). In marine systems, this idea has not been rigorously tested and remains an open question (see Ruiz et al., 2000). Studies have shown that reducing space competition and diversity can locally (at the small patch level) increase invasion success (Stachowicz et al., 1999). However, this is only a small-scale effect, and the degree to which human disturbance opens up free space in space-limited systems and whether this increases the likelihood of establishment and subsequent spread remains uncertain.

THE BEST-CASE SCENARIO FOR AN INVASION

The above sections suggest the conditions under which the probability of the invasion of a nonindigenous species would likely be maximized. High-quality propagules released in a retentive environment that closely matches that of the species' origin are more likely to survive and develop initial populations. Inocula with high genetic variability may also be favored. Further favoring establishment are biological characteristics such as being eurytopic and euryphagic, and life history characteristics such as direct development, asexual reproduction, and/or the ability to form resting stages. Once initial reproducing populations are formed, efficient dispersal capabilities combined with habitat connectivity may insure the development of large effective population sizes that would reduce inbreeding and the probability of extinction. Finally, significantly increasing the probability of establishment could be the absence of predators, parasites, and competitors.

While it is extremely difficult to quantitatively prioritize all of these variables that vary across species, space, and time, initial habitat compatibility and propagule retention are likely to be critical. The importance of the latter two variables underscore the relative role of inoculum density: very low inocula that match these two scenarios along with the other "best-case" criteria above could lead to highly successful invasions. Thus, one female crab with stored sperm introduced into a lagoon with highly restricted tidal exchange could produce an initial and even dense first-generation population of crabs that could then interbreed, and eventually exit the site that had exacted as an incubator. Conversely, of course, any best-case situation can quickly turn into invasion failure, such that propagules of the very highest quality (in terms of potential for settlement and survival) could arrive at a location where there would be instant mortality from any number of variable physical, chemical, or biological processes. In conclusion, the thread of inherent stochasticity winds inextricably through invasion dynamics.

CONCLUSIONS

Although a general assumption is that a decrease in the delivery of propagules of nonindigenous species—in terms of both release abundance and frequency—to a new region will result in a decrease in invasions, the *precise* nature of the response can vary enormously over species, time, and environments. In short, while inoculum density is a key component, it alone is not sufficient to explain the entire invasion process, which involves a large number of variables. These variables arise from a complex, reticulate mesh that weaves together the broad concept of propagule pressure (which itself involves many inter-related and often difficult-to-quantify phenomena and processes) with species traits and the multifaceted biological, ecological, physical, and chemical nature of the recipient environment, which changes over time. Inoculum density, the basis of proposed discharge standards, is thus but one of scores of variables that can and do influence invasion outcome. **Any method that attempts to predict invasion outcomes based upon only one factor in the multi-dimensional world of the invasion process is likely to suffer from a high level of uncertainty.**

REFERENCES

Behrens Yamada, S., B. R. Dumbauld, A. Kalin, C. E. Hunt, R. Figlar-Barnes, and A. Randall. 2005. Growth and persistence of a recent invader *Carcinus maenas* in estuaries of the northeastern Pacific. Biological Invasions 7:309–321.

Bender, E. A., T. J. Case, and M. E. Gilpin. 1984. Perturbation Experiments in Community Ecology: Theory and Practice. Ecology 65:1–13.

Blossey, B., and R. Nötzold. 1995. Evolution of increased competitive ability in invasive non-indigenous plants: a hypothesis. Journal of Ecology 83:887–889.

Briggs, J. C. 1974. Marine zoogeography. New York: McGraw-Hill Book Co.

Cadotte, M. W., S. M. McMahon, and T. Fukami, editors. 2006. Conceptual ecology and invasion biology: reciprocal approaches to nature. New York: Springer.

Callaway, R. M., and W. M. Ridenour. 2004. Novel weapons: a biochemically based hypothesis for invasive success and the evolution of increased competitive ability. Frontiers in Ecology and the Environment 2:436–433.

Carlton, J. T. 1985. Transoceanic and interoceanic dispersal of coastal marine organisms: the biology of ballast water. Oceanography and Marine Biology: An Annual Review 23:313–371.

Carlton, J. T. 1992. Introduced marine and estuarine mollusks of North America: an end-of-the-20th-century perspective. Journal of Shellfish Research 11:489–505.

Carlton, J. T., and A. N. Cohen. 1998. Periwinkle's progress: The Atlantic snail *Littorina saxatilis* (Mollusca: Gastropoda) establishes a colony on a Pacific shore. The Veliger 41:333–338.

Carlton, J. T. and E. A. Norse. 1993. Alien species. Pp. 130–136 In: Global Marine Biological Diversity. A Strategy for Building Conservation into Decision Making. E. A. Norse, editor. Washington, DC: Island Press. 383 pp.

Case, T. J. 1996. Global patterns of establishment and distribution of exotic birds. Biological Conservation 78:69–96.

Cohen, A. N., and J. T. Carlton. 1995. Biological Study. Nonindigenous Aquatic Species in a United States Estuary: A Case Study of the Biological Invasions of the San Francisco Bay and Delta. A Report for the United States Fish and Wildlife Service, Washington, DC, and The National Sea Grant College Program, Connecticut Sea Grant, NTIS Report Number PB96-166525.

Cohen, A. N., J. T. Carlton, and M. C. Fountain. 1995. Introduction, dispersal and potential impacts of the green crab *Carcinus maenas* in San Francisco Bay, California. Marine Biology 122:225–237.

Colautti, R. I. 2005. Are characteristics of introduced salmonid fishes biased by propagule pressure? Canadian Journal of Fisheries and Aquatic Sciences 62:950–959.

Crisp, D. J. 1958. The spread of *Elminius modestus* Darwin in north-west Europe. Journal of the Marine Biological Association of the United Kingdom 37:483–520.

Davis, M. A. 2009. Invasion biology. Oxford: Oxford University Press.

Ehrlich, P. R. 1986. Which animal will invade? Pp. 79-95 In: Ecology of Biological Invasions of North America and Hawaii. H. A. Mooney and J. A. Drake, editors. New York: Springer-Verlag.

Ekman, S. 1953. Zoogeography of the sea. London: Sidgwick and Jackson.

Elith, J., and J. R. Leathwick. 2009. Species distribution models: ecological explanation and prediction across space and time. Annual Review of Ecology and Systematics 40:677–697.

Elton, C. S. 1958. The ecology of invasions by animals and plants. London: Methuen.

Galil, B., and A. Zenetos. 2002. A sea change: exotics in the Eastern Mediterranean Sea. Pp. 325–336 In: Invasive aquatic species of Europe: distribution, impacts and management. Leppakoski E., S. Gollasch S. and S. Olenin, editors. Dordrecht, The Netherlands: Kluwer Academic Publishers.

Glasby, T. M., S. D. Connell, M. G. Holloway, and C. L. Hewitt. 2007. Nonindigenous biota on artificial structures: could habitat creation facilitate biological invasions? Marine Biology 151:887–895.

Gollasch, S. Lenz, J. Dammer, M. and H.G. Andres. 2000. Survival of tropical ballast water organisms during a cruise from the Indian Ocean to the North Sea. Journal of Plankton Research 22:923–937.

Goulletquer, P., G. Bachelet, P. G. Sauriau, P. G., and P. Noel. 2002. Open Atlantic coast of Europe. A century of introduced species into French waters. Pp. 276–290 In: Invasive aquatic species of Europe: distribution, impacts and management. Leppakoski E., S. Gollasch S. and S. Olenin, editors. Dordrecht, The Netherlands: Kluwer Academic Publishers.

Grosholz, E. D. 2005. Recent biological invasion may hasten invasional meltdown by accelerating historical introductions. Proceedings of the National Academy of Sciences 102:1088–1091.

Grosholz, E. D., L. A. Levin, C. Tyler, and C. Neira. 2009. Changes in community structure and ecosystem function following *Spartina alterniflora* invasion of Pacific estuaries. Pp. 23–40 In: Human impacts on salt marshes: a global perspective. B. Silliman, M. Bertness, and E. Grosholz, editors. Berkeley, CA: University of California Press. 432 pp.

Hunt, J., L. F. Bussiere, M. D. Jennions, and R. Brooks. 2004. What is genetic quality? Trends in Ecology and Evolution 19:329–333.

Johannesson, K. 1988. The paradox of Rockall: why is a brooking gastropod (*Littorina saxatilis*) more widespread than one having a planktonic larval dispersal stage (*L. littorea*)? Marine Biology 99:507–513.

Levin L. A., C. Neira, and E. D. Grosholz. 2006. Invasive cordgrass modifies wetland trophic function. Ecology 87:419–432.

Levinton, J. S. 2008. Marine Biology. Function, Biodiversity, Ecology. Third Edition. Oxford: Oxford University Press.

Lockwood, J., M. Hoopes, and M. Marchetti. 2006. Invasion ecology. New York: Wiley-Blackwell.

Lonsdale, W. M. 1999. Global patterns of plant invasions and the concept of invasibility. Ecology 80:1522–1536.

Losos, J. B., and R. E. Ricklefs, editors. 2009. The theory of island biogeography revisited. Princeton, NJ: Princeton University Press.

Mack, R. N., D. Simberloff, W. M. Lonsdale, H. Evans, M. Clout, and F. A. Bazzaz. 2000. Biotic invasions: causes, epidemiology, global consequences and control. Ecological Applications 10:689–710.

Neira, C., L. Levin, and E. Grosholz. 2005. Alteration of wetland ecosystems in San Francisco Bay by *Spartina* hybrid invasion. Marine Ecology Progress Series 292:111–126.

Neira, C., L. A. Levin, E. D. Grosholz, and G. Mendoza. 2007. Influence of invasive *Spartina* growth stages on associated macrofaunal communities. Biological Invasions 9:975–993.

Parker, J. D., D. E. Burkepile, and M. E. Hay. 2006. Opposing effects of native and exotic herbivores on plant invasions. Science 311:1459–1461.

Phillips, S. J., and M. Dudik. 2008. Modeling of species distributions with Maxent: new extensions and a comprehensive evaluation. Ecography 31:161–175.

Ricklefs, R. E., and G. Miller. 1999. Ecology. Fourth edition. San Francisco, CA: W. H. Freeman.

Roman, J. 2006. Diluting the founder effect: cryptic invasions expand a marine invader's range. Proceedings of the Royal Society of London, Biological Sciences 273:2453–2459.

Rudnick, D., T. Veldhuizen, R. Tullis, C. Culver, K. Hieb, and B. Tsukimura. 2005. A life history model for the San Francisco Estuary population of the Chinese mitten crab, *Eriocheir sinensis* (Decapoda : Grapsoidea). Biological Invasions 7:333–350.

Ruiz, G. M., P. Fofonoff, J. T. Carlton, M. J. Wonham, and A. H. Hines. 2000. Invasions of Coastal Marine Communities in North America: Apparent Patterns, Processes, and Biases. Ann. Rev. Ecol. Syst. 31:481–531.

Ruiz, G. M., and D. F. Reid. 2007. Current state of understanding about the effectiveness of ballast water exchange (BWE) in reducing aquatic nonindigenous species (ANS) introductions to the Great Lakes Basin and Chesapeake Bay, USA: synthesis and analysis of existing information. NOAA Technical Memorandum, Great Lakes Environmental Research Laboratory (GLERL) 142.

Sakai, A. K., F. W. Allendorf, J. S. Holt, D. M. Lodge, J. Molofsky, K. A. With, S. Baughman, R. J. Cabin, J. E. Cohen, N. C. Ellstrand, D. E. McCauley, P. O'Neil, I. M. Parker, J. N. Thompson, and S. G. Weller. 2001. The Population Biology of Invasive Species. Annual Review of Ecology and Systematics 32:305–332.

Saltonstall, K. 2002. Cryptic invasion by a non-native genotype of the common reed, *Phragmites australis*, into North America. Proceedings of the National Academy of Sciences 99:2445–2449.

Santagata, S., Z. R. Gasiūnaite, E. Verling, J. R. Cordell, K. Eason, J. S. Cohen, K. Bacela, G. Quilez-Badia, T. H. Johengen, D. F. Reid, and G. R. Ruiz. 2008. Effect of osmotic shock as a management strategy to reduce transfers of nonindigenous species among low-salinity ports by ships. Aquatic Invasions 3:61–76.

Sax, D. F., J. J. Stachowicz, and S. D. Gaines, editors. 2005. Species invasions: insights into ecology, evolution, and biogeography. Sunderland, MA: Sinauer Associates.

Seiden, J. M., C. Way, and R. B. Rivkin. 2010. Microbial hitchhikers: dynamics of bacterial populations in ballast water during a trans-Pacific voyage of a bulk carrier. Aquatic Invasions 5:13–22.

Simberloff, D. 1989. Which insect introductions and which fail? Pp. 61–75 In: Biological invasions: A global perspective. J. A. Drake, H. A. Mooney, F. di Castri, R. H. Groves, F. J. Kruger, M. Rejmanek, and M. Williamson, editors. New York: John Wiley and Sons.

Simberloff, D. 2006. Invasional meltdown 6 years later: important phenomenon, unfortunate metaphor, or both? Ecology Letters 9:912–919.

Sorte, C. J. B., S. L. Williams, and J. T. Carlton. 2010. Marine range shifts and species introductions: comparative spread rates and community impacts. Global Ecology and Biogeography 19:303–316.

Stachowicz, J. J., R. B. Whitlatch, and R. W. Osman. 1999. Species diversity and invasion resistance in a marine ecosystem. Science 286:1577–1579.

Stohlgren, T. J., D. Binkley, G. W. Chong, M. A. Kalkham, L. D. Schell, K. A. Bull, Y. Otsuki, G. Newman, M. Bashkin, and Y. Son. 1999. Exotic plant species invade hot spots of native plant diversity. Ecological Monographs 69:25–46.

Thiel, M., and P. A. Haye. 2006. The ecology of rafting in the marine environment. III. Biogeographical and evolutionary consequences. Oceanography and Marine Biology: An Annual Review 44:323–429.

Torchin, M. E., K. D. Lafferty, A. P. Dobson, V. J. McKenzie, and A. M. Kuris. 2003. Introduced species and their missing parasites. Nature 421:628–630.

Udvardy, M. D. F. 1975. A classification of the biogeographical provinces of the world. IUCN Occasional Paper 18, Switzerland.

Vasquez, E. A., E. P. Glenn, J. J. Brown, G. R. Guntenspergen, and S. G. Nelson. 2005. Salt tolerance underlies the cryptic invasion of North American salt marshes by an introduced haplotype of the common reed *Phragmites australis* (Poaceae). Marine Ecology Progress Series 298:1–8.

Verling, E., G. M. Ruiz, L. D. Smith, B. Galil, A. W. Miller, and K. R. Murphy. 2010. Supply-side invasion ecology: characterizing propagule pressure in coastal ecosystems. Proceedings of the Royal Society B 272:1249–1257.

Vermeij, G .J. 1991. When biotas meet: understanding biotic interchange. Science 253:1099–1104.

Vermeij, G. J. 1996. An agenda for invasion biology. Biological Conservation 78:3–9.

Williamson, M. 1996. Biological invasions. London: Chapman and Hall.

Wonham, M. J., J. T. Carlton, G. M. Ruiz and L. D. Smith. 2000. Fish and ships: relating dispersal frequency to success in biological invasions. Marine Biology 136:1111–1121.

4
Relationship between Propagule Pressure and Establishment Risk

Chapter 3 highlighted the many factors that influence invasion risk, namely (in broad terms) propagule pressure, species traits, abiotic environmental characteristics, and biotic interactions. Managing invasion risk by setting a discharge standard assumes that, despite these powerful modifying factors, organism density alone is a reasonable predictor of establishment probability. Consequently, this chapter examines the relationship between organism density and invasion risk, and considers how this relationship might help inform an organism-based discharge standard. Subsequent chapters examine other, non-modeling-based approaches to setting a discharge standard (Chapter 5) and evaluate the data requirements and limitations in estimating the relationship between invasion risk and propagule pressure, including uncertainty, variability over space and time, the relative merits of historical, survey, and experimental data, and the use of proxy variables (Chapter 6).

THE RISK–RELEASE RELATIONSHIP

Concepts and Terms

There are many definitions of invasion risk (e.g., Drake and Jerde, 2009). This report uses the term *invasion risk* interchangeably with *establishment probability* to refer to the chance that an introduced group of individuals establishes a self-maintaining population. In formal risk assessment, risk is defined a function of both exposure (the probability of a harmful event) and hazard (the effect of the harmful event). In this framework, invasion risk *could* be defined as a function of the probability of a species establishing (exposure) and its expected impact (hazard); however, such terminology is not used here. Rather, the term *invasion risk* is defined simply as the probability of establishment. For repeated introductions of invasive species, it is important to consider the time scale for establishment probability because, over infinite time, any invasion with a non-zero probability will eventually occur. The term *invasion rate* refers to the number of nonindigenous species that establish in a given region per unit time. It is straightforward to convert invasion rate to establishment probability.

A *propagule* is any biological material (such as particles, cells, spores, eggs, larvae, and mature organisms) that is or may become a mature organism. *Propagule pressure* is a general term expressing the quantity, quality, and frequency with which propagules are introduced to a given location. As discussed in Chapter 3, propagule pressure is a function of a suite of variables reflecting the nature of the species and the transport vector. The remainder of this chapter, however, focuses on the quantity of propagules alone.

In the context of ballast water, it is useful to distinguish two measures of propagule quantity. Following Minton et al. (2005), *inoculum density* is defined as the density of organisms in released ballast water. Inoculum density (denoted as D_I in equations) is given simply as the total number of organisms in the inoculum (N_I) divided by the inoculum, or ballast water, volume (V_I):

$$D_I = N_I / V_I.$$

The *initial population size* is the initial number of organisms released into the environment in a given location at a given time, i.e., the inoculum number (N_I). As these organisms will tend to spread out into their new habitat, their density in the environment (D_E) is, in the simplest formulation, given as the number of organisms released (N_I) divided by the volume of water in the environment (V_E):

$$D_E = N_I / V_E.$$

It is important to recognize that it is the inoculum density (D_I) that is subject to a ballast water discharge standard. However, both the initial number of organisms N_I (conventionally denoted N_0 in population modeling) and their density in the new environment D_E are expected to affect establishment probability. In other words, the variable to be managed to reduce invasion risk is clearly distinct from the variable that is typically used in predicting invasion risk. This disconnect is the central conceptual challenge in converting empirical and theoretical results in population establishment to an operational discharge standard (see section below).

The relationship between invasion risk and propagule pressure is the *risk–release relationship*. Understanding the risk–release relationship is essential to predicting and comparing the invasion risk associated with different discharge standards. However, understanding this relationship is not a straightforward proposition. It is easiest to define, model, and estimate this relationship for single species, focusing on the relationship between the number of individuals released at a given time in a given location, and the probability of that population establishing. For larger-scale analyses of invasions by multiple species, the relationship becomes less easy to define, model, and measure. In general, however, it can be thought of as the number of established species as a function of the number of released species, organisms, or combination thereof, over a given time period. In the case of ballast water, the central but somewhat indirect risk–release relationship is the number of species that establish as a function of the large-scale release of a varying number of varying species at varying densities.

The Hypothetical Risk–Release Relationship

In general, the relationship between invasion risk and propagule pressure is expected to be positive, although its shape is unknown. A priori, it might take any of a number of standard shapes including linear, exponential, hyperbolic, and sigmoid (Ruiz and Carlton, 2003) as shown in Figure 4-1. The shape of the risk–release relationship has important implications for managing invasion risk.

If the relationship were linear, then a given reduction in release density would always lead to a proportional reduction in invasion risk (Figure 4-2A). If the relationship were nonlinear, with one or more inflection or slope-balance points, then interesting management thresholds would emerge (Figure 4-2B). If the relationship were exponential, a reduction from high to moderate release density leads to a much greater reduction in invasion risk than a similar-sized reduction from moderate to low release density. The opposite would hold for

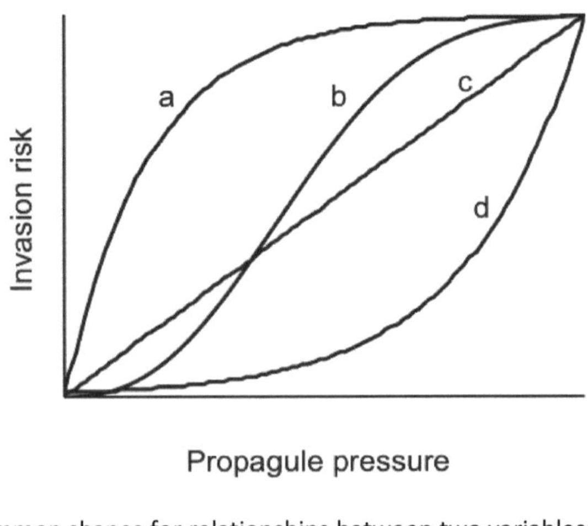

FIGURE 4-1 Common shapes for relationships between two variables: (a) hyperbolic, (b) sigmoid, (c) linear, (d) exponential (both axes linear). A priori, any of these could represent the relationship between invasion risk (probability of a species establishing) and propagule pressure (e.g., number of individuals released). SOURCE: Adapted, with permission, from Ruiz and Carlton (2003). ©2003 by Island Press.

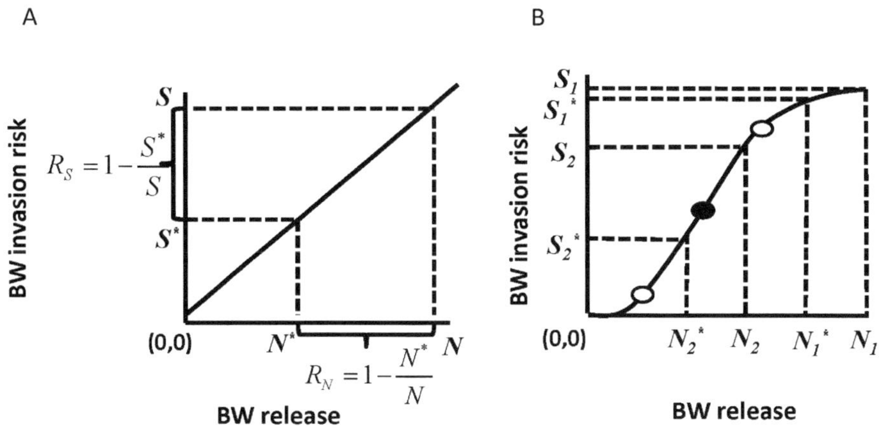

FIGURE 4-2 Conceptual application of a risk–release relationship to inform ballast water organism discharge standards. *S* is the observed risk of ballast-mediated species invasion; *N* is the observed number of organisms released. (Axis units depend on whether the model represents a single species or multiple species.) S^* is the target invasion risk; N^* is the corresponding target release value. Assuming a robust risk–release relationship, reducing the ballast water release by the proportion R_N is predicted to reduce the invasion risk by the proportion R_S. (A) Under the assumption of a linear risk-release relationship, a given reduction in the release rate is predicted to give the same proportional reduction in invasion rate (i.e., $R_N = R_S$). (B) Under the assumption of a sigmoid relationship, the same reduction in release ($R_{N1} = 1 - [N^*_1/N_1]$ = $R_{N2} = 1 - [N^*_2/N_2]$) is predicted to give a much lesser ($R_{S1} = 1 - [S^*_1/S_1]$) or a much greater ($R_{S2} = 1 - [S^*_2/S_2]$) reduction in invasion risk, depending on the range over which R_N occurs. Solid dot indicates the inflection point around which the greatest reduction in risk is obtained for the least reduction in release. Open dots indicate the points at which the slope passes through a 45° angle: outside these bounds, increasingly less risk reduction is obtained for the same release reduction. Panel (A) provides a graphical illustration of the multi-species linear model proposed by Cohen (2005, 2010) and Reusser (2010), where the y-axis is the invasion risk (characterized as invasion rate, the number of species established per unit time) and the x-axis is organism release (characterized as release rate, the total number organisms per unit time). Reusser (2010) defined as the *per capita invasion risk* the number of introduced species that establish per organism released, i.e., the slope of the straight-line relationship. Cohen (2005, 2010) defined as the *reduction factor* the equal proportional reductions in release and risk from current to target levels.

a hyperbolic relationship. For an S-shaped curve, a reduction over the middle range of release densities would to the greatest reduction in invasion risk (Figure 4-2B). Because of qualitative change in the risk–release relationship at the inflection point and at the slope-balance points (where the tangent to the curve passes through 45°), non-linear relationships present influential management thresholds.

Theory tells us that for a single population, only two of these four shapes—the hyperbolic and sigmoid curves—can represent the overall risk–release relationship (see Box 4-1). The combined curves of multiple populations would also be expected to be nonlinear. Thus, it is expected that there should be at least one threshold in the risk–release relationship that could in principle prove useful in informing discharge standards.

Despite understanding that the overall theoretical shape is hyperbolic or sigmoid for a single species, it is possible for a given set of risk–release data to be better characterized by a linear or even an altogether different model. This apparent discrepancy could emerge for two main reasons. First, there may be insufficient data points to support a curved line over a straight line. This difficulty will be exacerbated as the true slope decreases (for example, the lower-left or upper-right ends of the hyperbolic or sigmoid curves). Second, any underlying theoretical risk–release relationship may be swamped out by other more important sources of variation that affect establishment probability (see Chapter 3), such that it cannot be recovered from the data.

To quantitatively predict the effects of a discharge standard on invasion risk, and to compare the risk associated with different discharge standards, it is essential to understand the shape and strength of the risk–release relationship. The following section reviews a range of approaches that have been taken to fitting risk–release curves to empirical data.

Modeling the Risk–Release Relationship

An ideal analysis of the risk–release relationship would involve developing and testing a suite of candidate theoretical models, collecting multiple rigorous empirical datasets, and comparing the fit of the models to the data to determine (1) which model best captures the risk–release relationship and (2) how strong this relationship is relative to other potential explanatory variables.

Several approaches have been taken to modeling the risk–release relationship, categorized in Table 4-1 along two axes. First, models can range from descriptive models that simply represent the shape of the relationship to mechanistic models that define the processes generating the relationship (e.g., Drake and Jerde, 2009). These are not mutually exclusive categories—a given model may include both mechanistic and descriptive components—but at their extremes they represent very different modeling philosophies and goals, and they define a useful spectrum for organizing modeling approaches. Second, Table 4-1 distinguishes models that focus on the establishment of a single species vs.

TABLE 4-1 Approaches Taken to Modeling the Relationship Between Invasion Risk and Propagule Pressure

		Scale	
Type	*Sample structure*[1]	*Single species examples*	*Multiple species examples*
Descriptive	Statistical Logistic regression: $\ln\left(\frac{y}{1-y}\right)=b_0+b_1x$ Functional form Species-area curve: $y=cx^z$	Memmott et al. (2005) Jongejans et al. (2007) Bertolino (2009) Drake and Jerde (2009)	Lonsdale (1999) Levine and D'Antonio (2003) Ricciardi (2006) Cohen (2005, 2010)* Reusser (2010)*
Mechanistic	Probabilistic $p_E=1-e^{-\alpha N}$ Dynamic demographic $\frac{dN}{dt}=rN$	Shea and Possingham (2000) Leung et al. (2004) Memmott et al. (2005) Drake et al. (2006) USCG (2008)* Drake and Jerde (2009) Jerde et al. (2009) Bailey et al. (2009) Kramer and Drake (2010)	Drake et al. (2005)* Costello et al. (2007)* USCG (2008)* See text

Notes: Models are categorized as being on the descriptive or the mechanistic end of a spectrum, and as representing single or multiple invading species. Sample structures show simple, generic forms of these model types. Most of these approaches have been widely developed and implemented throughout the biological literature; only a few recent examples, further discussed in the text, are listed here. Some studies illustrate more than one modeling approach.

[1] Parameters: *y*, dependent variable (invasion risk); *x*, independent variable (organism discharge or proxy variable); b_0, b_1, *c*, *z*, shape parameters; p_E, population establishment probability; α, ln(individual establishment probability); *N*, number of individuals; *r*, per capita population growth rate.

* Proposed application of modeling approach to ballast water management.

that of multiple species. The following sections highlight illustrative examples of each approach and outline their key advantages and disadvantages. Since population establishment theory applies across all species, habitats, and vectors, many of the given examples fall outside the immediate realm of ballast-mediated invasions. Nevertheless, the approaches illustrated are readily applicable to the risk–release relationship for ballast water.

SINGLE-SPECIES MODELS

It is informative to examine the risk–release relationship at the scale of a single species, for two main reasons. First, this approach allows examination of invasion scenarios for certain model species, such as fast growing, high impact, or commonly released invaders, which could be used to obtain upper bounds for discharge standards under best-case (for invasion) scenarios. Second, it allows for clarification of model structures and assumptions before scaling up to the more realistic scenario of multi-species releases. The primary disadvantage of the single-species approach, in the context of managing ballast water, is that it does not represent the reality of the simultaneous and continuous release of many species from ballast water. The greatest challenge in this approach is in converting experimental and theoretical results premised on N_0 to a discharge standard applicable to D_I.

A ballast-mediated invasion may be expected to begin from the introduction of relatively few individuals. Three factors are particularly relevant to small population dynamics—demographic stochasticity, positive density dependence, and the spatial environment—and must be considered in developing an effective risk–release model. Their net effect can be captured by a descriptive model; their individual effects can be tested by incorporating them explicitly into a mechanistic model.

Demographic stochasticity is the natural variability in individual survival and reproduction that occurs in populations of any size, but that in small populations can lead to large fluctuations in population growth rate. On average, demographic stochasticity makes extinction more likely than the equivalent deterministic model would predict; on the other hand, in a given realization, stochasticity can lead to establishment when a deterministic model would otherwise predict extinction (Morris and Doak, 2002; Drake, 2004; Andersen, 2005; Drake et al., 2006; Jerde et al., 2009). Demographic stochasticity is expected to lower the risk–release curve at low density.

Positive density dependence, or *Allee effects*, is the intuitively logical notion that as organism density decreases, individuals may suffer increasing difficulty finding mates or foraging. Population growth rate would thus be expected decline at low density, rather than to increase as per an exponential growth model (Allee, 1931; Courchamp et al., 2009). Thus, across the range of low initial densities, it takes a higher density to achieve the same invasion risk when Allee effects are at work than when they are not. Allee effects lower the risk–release curve at low density, leading to a characteristic sigmoid curve (see Curve B in Figure 4-1). Allee effects are logically appealing and have been applied to models of both sexually reproducing and parthenogenetic organisms (Drake, 2004). However, they have proved elusive to document empirically (see discussion and examples in Dennis, 2002; Morris and Doak, 2002; Leung et al., 2004; Drake et al., 2006; Courchamp et al., 2009; Jerde et al., 2009; Kramer and Drake, 2010).

The third feature is the effect of the *spatial environment* on population density. One of the greatest challenges in population modeling is that organisms released into an environment will tend to spread out, through both passive and active dispersal. Depending on the relative scales of dispersal and population growth, the *effective* initial population size may be very much lower than the original number of organisms released. This reduction in density will presumably exacerbate the impacts of demographic stochasticity and Allee effects. Thus, in general it is expected that both individual and population establishment probabilities to be much lower in the wild than in contained laboratory experiments. (It is possible, of course, that hydrodynamic features or aggregative behavior will have the opposite effect, tending to concentrate organisms in a locale and enhancing their chances of establishment; see Chapter 3).

Since establishment probability seems generally likely to be dominated by the dynamics of small populations, the most rigorous modeling approach is to construct models that incorporate (or can phenomenologically reflect the effects of) demographic stochasticity, Allee effects, and their modification by dispersal, and to allow the empirical data to indicate on a case-by-case basis the importance of these additional features. The following examples consider none, some, or all of these features.

Descriptive Models

Descriptive models, such as regression and similar statistical techniques, offer a phenomenological characterization of the risk–release relationship. That is, they can be formulated and parameterized without having to understand or specify the underlying mechanisms by which the independent variables explain the dependent variable (Drake and Jerde, 2009). As a result, one's confidence in their predictive ability is limited. Familiar descriptive models include statistical models such as regression and functional forms such as species-area curves and behavioral responses.

These models have been applied to the results of both experimental and historical survey data. In a simple example, Drake and Jerde (2009) fit a spline, or a series of local regressions, to establishment probability as a function of propagule pressure in the scentless chamomile (*Matricaria perforata*) (Figure 4-3A). In this case, establishment was defined as survival simply from seed to flowering, but the same method could be applied to a longer-term study of population establishment. The same data were also fit with a probabilistic model (see later section). In a slightly more complex field experiment, Jongejans et al. (2007)

estimated the establishment probability (defined as persistence over six years) of the European thistle (*Carduus acanthoides*) as a function of propagule pressure and native plant biomass. Using a generalized linear model, they found that establishment probability increased significantly with higher initial seed number, and tended to increase with reduced native biomass; together these variables accounted for 37 percent of the variation (Figure 4-3B). Bertolino (2009) modeled the success of global historical squirrel introductions (defined as persistence to the present day of populations introduced over a >100-year period) as a function of propagule pressure, environmental matching, native diversity, and the invaders' biogeographical origin. For the genus *Sciurus* (squirrels), a logistic regression fit to the initial number of individuals alone explained 55 percent of the variation in establishment probability (Figure 4-3C).

Statistical models can be made increasingly complex by adding ever more independent variables, and have been used to describe invasion risk over a spatial domain using species distribution (environmental niche) modeling (Peterson and Vieglais, 2001; Herborg et al., 2007, 2009; Dullinger et al., 2009).

Mechanistic Models

In contrast to descriptive statistical models, mechanistic models represent invasion establishment as a function of parameters that have a readily defined biological meaning. Whereas statistical models describe a relationship only over the range of data to which they are fit, mechanistic models are presumed to extrapolate well over the entire biologically realistic parameter space. Furthermore, descriptive models allow one to investigate the shape of a relationship, while mechanistic models force the user to specify the processes driving the relationship and to link causative variables explicitly.

It is useful to distinguish two general classes of mechanistic models: probabilistic statements and dynamic, demographic models. In their simpler forms, these two model classes possess different mathematical structures and require different data to parameterize and validate. In more complex models of population establishment, this distinction blurs and a given model may incorporate elements of both classes (e.g., Jerde and Lewis, 2007; Leung and Mandrak, 2007; Jerde et al., 2009).

Probabilistic Models

In the context of the risk–release relationship, the *probabilistic models* considered here are composed of probability statements beginning with the probability of an individual's establishment probability and scaling up to a population level. A probabilistic model is written immediately in terms of its solution, namely, in terms of population establishment probability. When the model's constituent probabilities are represented as frequency distributions rather than as

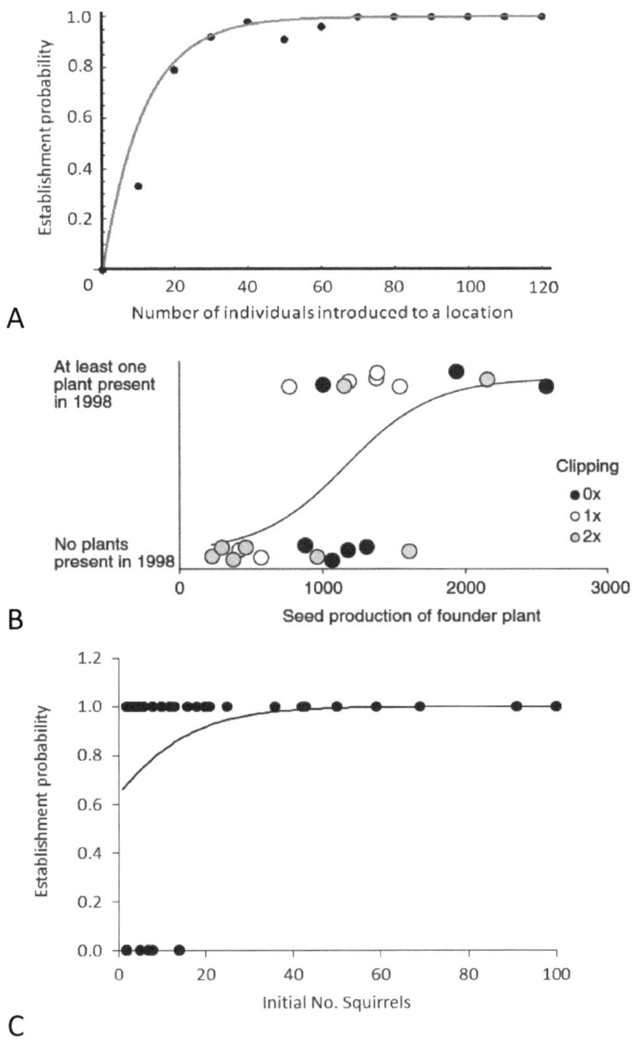

FIGURE 4-3 Single-species risk–release relationships obtained from descriptive models. (A) Spline fit to short-term establishment probability of scentless chamomile (*Matricaria perforata*) (Drake and Jerde, 2009). (B) Generalized linear model fit to establishment outcome (success or failure) for European thistle (*Carduus acanthoides*) across three levels of native plant biomass reduced by clipping (Jongejans et al., 2007). (C) Logistic regression fit to establishment probability of squirrel *Sciurus*. Replotted from Bertolino (2009) with data generously provide by S. Bertolino. SOURCES: (A) Reprinted, with permission from, Drake and Jerde (2009). © 2009 by Oxford University Press. (B) Reprinted, with permission, from Jongejans et al. (2009). © by Springer. (C) Reprinted, with permission, from Bertolino (2009). © 2009 by John Wiley and Sons.

fixed points, it is known as a hierarchical probability model (HPM). The HPM approach to representing stochastic events has been extensively applied to medical, engineering, and ecological problems, and it allows the explicit incorporation and analysis of uncertainty (Dennis and Lele, 2009; Ponciano et al., 2009). Parameterizing and validating probabilistic models requires comparatively simple data: the outcome, as success or failure, of a series of introductions inoculated over a range of initial organism numbers. As will be discussed later, probabilistic models can readily be expanded to represent multiple species and environmental conditions that have different associated probabilities of establishment.

Probabilistic models of population establishment have been developed to serve as the basis for a metapopulation model of biocontrol release (Shea and Possingham, 2000) and for a gravity model of zebra mussel spread (Leung et al., 2004). In their simplest, non-spatial form, they contain a sole parameter—the probability of a single individual producing an established population. This value is then scaled up to obtain the probability of a group of individuals leading to an established population (Shea and Possingham, 2000; Leung et al., 2004, Leung and Mandrak, 2007; Jerde and Lewis, 2007; Jerde et al., 2009). The diffusion approximation to exponential growth shares the same core probability structure, and has likewise been used to model population establishment (Drake et al., 2006; Bailey et al., 2009).

The basic construction of a simple probabilistic model is outlined in Box 4-1, and its application is illustrated in several examples below.

To implement these probabilistic models, studies have examined population establishment over a range of initial population sizes either from directed experiments or from descriptive population data. Memmott et al. (2005) fit both a logistic regression and a probabilistic model to the success after six years of biocontrol insect releases (*Arytainilla spartiophila*; Figure 4-5A). Drake and Jerde (2009) used short-term data for the success of the prairie weed scentless chamomile from seed to flowering (*Matricaria perforata*; Jerde and Lewis, 2007) to scale up to a population-level establishment model. This model was subsequently extended by Jerde et al. (2009) to incorporate mate-finding limitations that led to a biologically driven Allee effect, and it was used to predict invasion risk for Chinese mitten crabs (*Eriocheir sinensis*) and apple snails (*Pomacea canaliculata*) (Jerde et al., 2009; Figure 4-5B). Drake et al. (2006) used descriptive population growth data of the spiny waterflea *Bythotrephes longimanus* in three lakes over four years, and Bailey et al. (2009) conducted 100-day mesocosm studies of a variety of cladocerans, to parameterize diffusion approximations to exponential growth models (Figure 4-5C,D).

Probabilistic models have also been constructed to investigate the accumulation of nonindigenous species over time (Solow and Costello, 2004; Wonham and Pachepsky, 2005); this type of model has been extended to examine the relationship between propagule pressure and invasion risk (Costello et al., 2007; see later section). Basic probabilistic models can represent the invasion outcome alone, without necessarily representing the population dynamics leading to success or failure. In contrast, dynamic demographic models directly represent population dynamics in order to predict invasion outcomes.

Dynamic Demographic Models

Dynamic models of population growth are written as a system of one or more discrete or differential equations whose solution gives the population size at a given time. Familiar examples include exponential (geometric) and logistic growth, and Leslie and Lefkovitch matrix models. The parameters in these models represent the demographic rates or probabilities of birth, death, growth, and reproduction. Dynamic demographic models do not predict establishment probability directly. Instead, a number of stochastic simulations must be generated for each initial population size. A subsequent model, either descriptive or probabilistic, must then be fit to the outcome of the simulations. Parameterizing and validating this kind of stochastic dynamic model requires estimating the distribution of each demographic parameter (e.g., birth rate, death rate). In their simpler forms, demographic models require more data to parameterize and validate than do probabilistic models.

Dynamic demographic models serve as the basis for the population viability analysis (PVA) of declining species (Morris and Doak, 2002), an approach that has been applied more recently to the analysis of establishing invaders (Bartell and Nair, 2004; Neubert and Parker, 2004; Andersen, 2005), and that has been proposed for use in comparing ballast water discharge standards (USCG, 2008). It is useful to recall that the goals and outcome of population viability analysis are not the exclusive domain of demographic models. Although the term PVA typically refers to the analysis of these models, any modeled population may be subjected to an analysis of its viability; indeed, such analysis is inherent in the construction of a probabilistic establishment model. Traditionally, PVA involves dynamic demographic models that are matrix-based and use age- or stage-specific dynamic rates to estimate population growth and hence viability. These models use dynamic information for each stage including growth, survival, and reproduction to estimate population growth rate (Caswell, 1989).

The most basic dynamic demographic models are count-based and assume no variation among individuals in the population. These count-based models make several assumptions including that the mean and variance of population growth remain constant, no density dependence, dynamic stochasticity, etc. (Morris and Doak, 2002). However, more complex count-based models can incorporate positive and negative density dependence, Allee effects, stochasticity, spatial structure, etc. The more realistic dynamic demographic models explicitly use different dynamic parameters for each age or stage class in the population (see example from Kramer and Drake, 2010, below). Although various kinds of stochasticity and autocorrelation in vital rates can be includedthrough simulation methods, these models have their own restrictive assumptions of time invariance, stable age or stage distribution, etc. (Morris and Doak, 2002).

BOX 4-1
Probabilistic Model Framework

This box illustrates the development of the simple probabilistic model in Leung et al. (2004) and shows how it provides insight into the overall shape of the risk–release relationship. It begins with a simple probability statement in which N is the number of propagules released and p is the individual establishment probability of each propagule. In this case, $1\text{-}p$ is the probability of a single propagule failing to establish, and $(1\text{-}p)^N$ is the cumulative probability of all N individuals failing to establish. The probability of the population establishing, p_E, is therefore

$$p_E = 1-(1-p)^N \qquad (4\text{-}1)$$

It is mathematically convenient to define an additional parameter, $\alpha = -\ln(1\text{-}p)$, which allows us to rewrite (4-1) synonymously as

$$p_E = 1 - e^{-\alpha N} \qquad (4\text{-}2)$$

For this equation (4-1, 4-2), the shape of the risk–release relationship between p_E and N is hyperbolic, asymptoting towards $p_E = 1$ (Figure 4-4a). This must be the case because even if the individual establishment probability (p) is low, the total probability p_E increases inexorably to one as more and more individuals are released (Leung et al., 2004).

This model can be extended to incorporate the negative density dependence, or Allee effects, that may be expected to reduce p_E at low values of N (Leung et al., 2004; Jerde et al., 2009). The resulting sigmoid shape (Figure 4-4b) can be produced by adding the shape coefficient $c > 1$ to equation 4-2 (which follows the cumulative Weibull distribution), giving

$$p_E = 1 - e^{-\alpha N^c} \qquad (4\text{-}3)$$

Equation (4-2) is the special case of (4-3) where $c = 1$ and there is no strong Allee effect (Leung et al., 2004). Thus, the biological meaning of c can be interpreted as an Allee parameter. However, it should be noted that if Allee effects are to be considered, both α and c need to be fit simultaneously to describe a biological system.

These equations illustrate that simple probability statements combined with basic principles of population growth reveal two candidate shapes for the overall p_E vs. N curve: hyperbolic or sigmoid. However, the shape of the curve for a given empirical dataset over a limited parameter space may appear linear, particularly for high (both curves) and low (sigmoid curve) values of N. Short sections of the sigmoid curve could also appear exponential (left-hand side of curve) or hyperbolic (right-hand side).

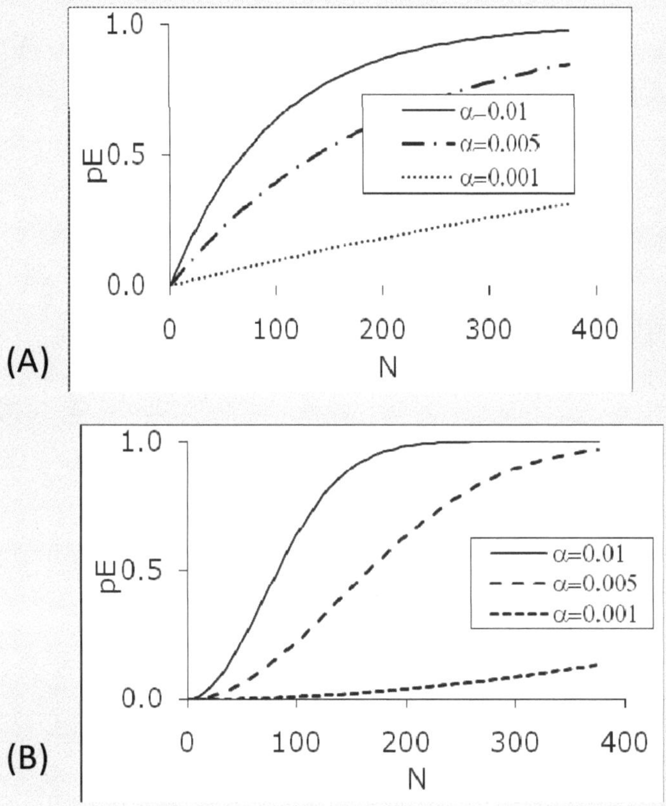

FIGURE 4-4 Predicted relationship between p_E and N given by equation 4-3. (A) Hyperbolic shape with no Allee effects (c = 1). (B) Sigmoid shape characteristic of an Allee effect (c > 1). In both panels, the upper, middle, and lower curves are for $p \approx \alpha$ = 0.01, 0.005, and 0.001.

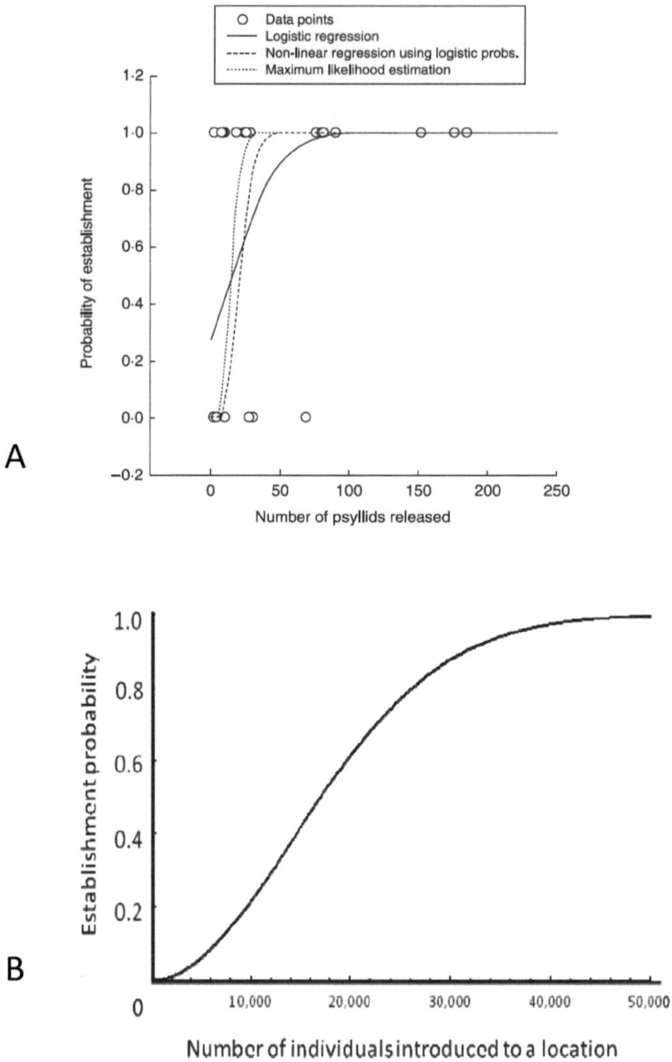
Data points
Logistic regression
Non-linear regression using logistic probs.
Maximum likelihood estimation
Probability of establishment
Number of psyllids released
A
Establishment probability
Number of individuals introduced to a location
B

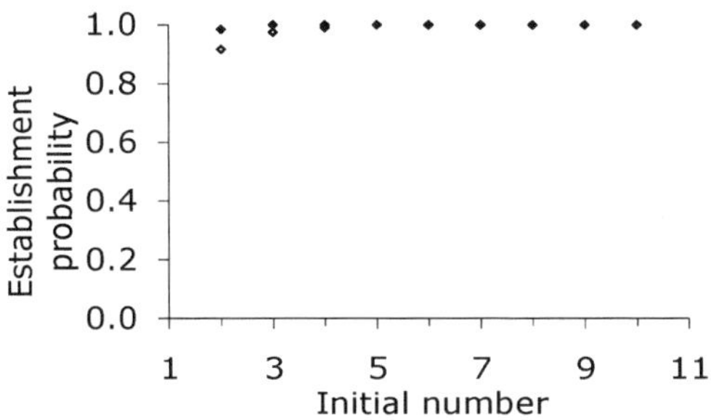

C

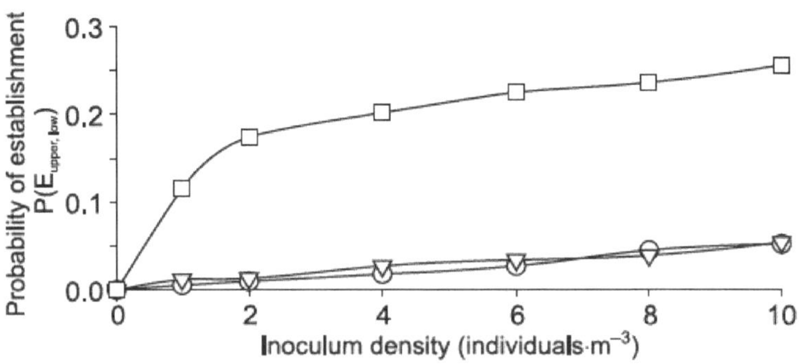

D

FIGURE 4-5 Continued Single-species risk–release relationships predicted by probabilistic models for the establishment probability of (A) the psyllid *Arytainilla spartiophila* (Memmott et al., 2005), (B) Chinese mitten crab (*Eriocheir sinensis*) (Jerde et al., 2009), (C) spiny waterflea (*Bythotrephes longimanus*) with demographic (open dots) and environmental (solid dots) stochasticity (redrawn from Drake et al., 2006), as a function of the initial number of organisms, and (D) three cladocerans (*Bosmina* spp., circles; *Bosmina coregoni*, triangles; *Daphnia retrocurva*, squares) as a function of the initial organism density (Bailey et al., 2009). SOURCES: (A) Reprinted, with permission, from Memmott et al. (2005). © 2005 by John Wiley and Sons. (B) Reprinted, with permission, from Jerde et al. (2009). © 2009 by The University of Chicago Press. (C) Reprinted, with permission, from Drake et al. (2006). © 2006 by Springer. (D) Reprinted, with permission, from Bailey et al. (2009). © 2009 by NRC Research Press.

Dynamic demographic models have been developed for a tremendous variety of plant and animal species. The following examples illustrate their application to predicting the establishment probability of introduced or re-introduced species. Wood et al. (2007) used life history data to parameterize an individual-based simulation model of tree squirrel re-introductions, and predicted the proportion of populations above a threshold abundance (Figure 4-6A). Other models represent the establishment of age- or stage-structured populations (Parker, 2000; Barry and Levings, 2002; Kramer and Drake, 2010) (Figure 4-6B). Kramer and Drake (2010) used experimental laboratory results to parameterize a demographic model of the cladoceran *Daphnia magna*, and found that increasing predation shifted the risk–release relationship from a hyperbolic to the sigmoid shape characteristic of Allee effects (Box 4-2).

Once a group of organisms is released, they will disperse through advection and locomotion. These may lead to a net aggregation or dispersal. The effects of dispersal on population establishment have been explored in considerable detail by extending demographic models to a reaction–diffusion framework and its extensions (Skellam, 1951; Shigesada and Kawasaki, 1997; Lubina and Levin, 1988; Neubert and Parker, 2004; Lewis et al., 2005; Hastings et al., 2005). These models have been used to explore the persistence and spread of aquatic and marine species (Drake et al., 2005; Pachepsky et al., 2005; Lutscher et al., 2007, 2010; Dunstan and Bax, 2007; reviewed for marine invasions by Wonham and Lewis, 2009), but in general have not been used to predict risk–release relationships. For the application of a reaction–diffusion model to a multi-species scenario, see Drake et al. (2005).

Obtaining a Discharge Standard from Single Species Models

A single-species model of the risk–release relationship could provide insight into discharge standards in two main ways: to illustrate a best-case scenario, and to serve as a building block for multi-species models. To illustrate a best-case scenario, a model could be constructed and parameterized for fast-growing, high-impact, or commonly released species. Invasion risk could then be predicted under the assumption that all ballasted organisms belonged to this species, and were released under optimal conditions. This approach would lead to a conservative discharge standard.

The greatest difficulty in developing a discharge standard from a single-species model is that these models are generally constructed to represent a one-time introduction of a known initial number of individuals. However, ballast water discharge is a repeated event, which will tend to increase invasion risk, and the organisms may rapidly be redistributed in the physical environment, which may immediately alter the effective initial number of individuals with the

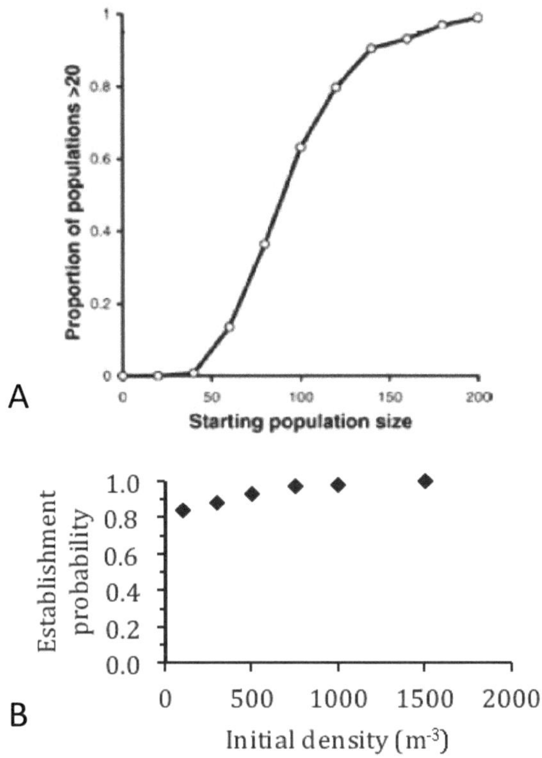

FIGURE 4-6 Single-species risk–release relationships obtained from dynamic demographic simulation models. (A) Predicted proportion of populations exceeding 20 individuals after 100 years from a population growth model of tree squirrels (Wood et al., 2007). (B) Predicted establishment probability from a population growth model of the copepod *Pseudodiaptomus marinus* (plotted from data in Barry and Levings, 2002). (A) Reprinted, with permission, from Wood et al. (2007). © 2007 by American Society of Mammalogists.

BOX 4-2
Dynamic Demographic Model Framework

This box illustrates the development of a dynamic demographic population model, following that formulated by Kramer and Drake (2010). The model framework begins with the standard continuous time equation for a homogeneous population of size N growing as a function of the difference between the birth rate (β) and the death rate (μ). The population growth rate is given as

$$\frac{dN}{dt} = \beta N - \mu N \quad (4\text{-}4)$$

To examine the effects of predation, an additional mortality function, $g(N)$, was added to represent a standard predation type II functional response, such that:

$$g(N) = \frac{P\alpha N}{(1+\alpha T_h N)} \quad (4\text{-}5)$$

where P is the number of predators, α is the attack rate, and T_h is the handling time, giving the population growth rate:

$$\frac{dN}{dt} = \beta N - \mu N - g(N) \quad (4\text{-}6)$$

The model was extended to represent two size classes, juveniles (J) and adults (A), where the juveniles are produced by adults at rate β and mature to adults at rate δ, and predation is a function of total population size:

$$\begin{aligned} \frac{dJ}{dt} &= \beta A - \mu J - g(J+A) - \delta J \\ \frac{dA}{dt} &= \delta J - \mu A - g(J+A) \end{aligned} \quad (4\text{-}7)$$

This model was parameterized from laboratory experiments with the cladoceran *Daphnia magna* and a non-visual ambush predator, larvae of the midge *Chaoborus trivittatus*. Stochastic model simulations showed that predation induced a sigmoid risk–release relationship, compared to the hyperbolic curve predicted in the absence of predation. In other words, predation induced an Allee effect in this system (Figure 4-7).

box continues next page

BOX 4-2 Continued

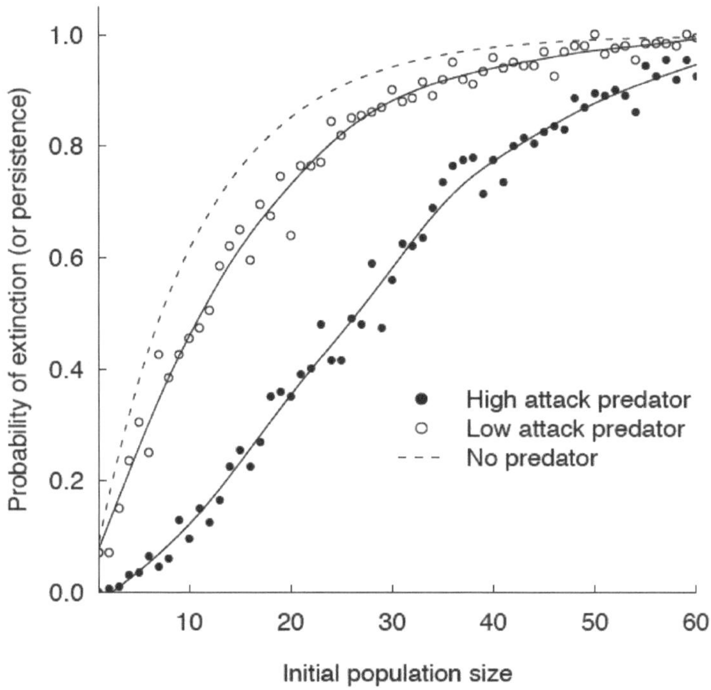

FIGURE 4-7 Predicted single-species risk–release relationship obtained from simulations of a dynamic demographic model. Theoretical (dashed line) and simulated (points with fitted solid-line spline) predictions of establishment probability for a *Daphnia magna* population as a function of initial populations size under different predation levels. SOURCE: Adapted from Kramer and Drake (2010); courtesy J. Drake. Reprinted, with permission, from Drake (2010). © 2010 by John Wiley and Sons.

potential to establish. The closer together small releases occur in space and time, the more they will approximate a single large release with a correspondingly higher establishment probability (analogous to the rescue effect in metapopulation dynamics, Gotelli, 1991). Theoretical studies have demonstrated that, due to environmental stochasticity, the likelihood of success of multiple arrivals at a single entry point is higher than that for simultaneous arrivals at multiple sites (Haccou and Iwasa, 1996). In a homogeneous environment, organisms will disperse and the effective initial population size will rapidly decrease; an advective environment and intraspecific behavior may either enhance or counter this effect. The mathematical framework of a single-species model could readily be modified to analyze the risk associated with multiple repeated inocula that are dispersed or concentrated in the local environment and to determine an adjusted discharge target.

In summary, both descriptive and mechanistic models have been developed to examine the risk–release relationship for single species. This relationship can reasonably easily be defined and parameterized for a one-time release under controlled laboratory or field conditions. It is somewhat more difficult to define and parameterize a model that would represent repeated releases in an advective environment, making it challenging to scale up to a discharge standard. The models could be useful either for setting a discharge standard based on a best-case species, or for developing modeling frameworks that would help inform multi-species scenarios.

MULTI-SPECIES APPROACHES

Broad-scale vectors like ballast water (Carlton and Geller, 1993; Smith et al., 1999), shipping containers (Suarez et al., 2005), or commercial imports (Copp et al., 2007; Dehnen-Schmutz et al., 2007) repeatedly release assemblages of tens to hundreds of species into the environment, of which only a small subset establish successfully. To model the risk–release relationship at this scale requires both risk and release data spanning the same large spatial, temporal, and taxonomic scales. At present, however, there are only loosely corresponding empirical estimates of risk and release (see Table 4-2). Before proceeding with multi-species examples, this section considers the nature of the available data for both invasion risk and organism release, and the resulting constraints on model construction and interpretation.

For **invasion risk**, there are historical records of the invaders that have accumulated in various ports over the past decades (see Table 4-2). These invasion records are characterized by considerable uncertainty stemming from incomplete collections, the cryptogenic nature of many species, the taxonomic bias of field samples, and the uncertainty associated with ascribing a given invasion to ballast transport over other candidate vectors (Chapman and Carlton, 1991; Ruiz et al., 2000; Costello et al., 2007; Fitzpatrick et al., 2009; Jerde and Bossenbroek, 2009). These are standard sampling difficulties that plague any assessment of nonindigenous species, and are not unique to the problem of ballast water management. The consequences, however, are that the best empirical estimates of invasion rate and risk are nevertheless incomplete and uncertain. Furthermore and more crucially, there are no good estimates of the scale of that uncertainty.

For **organism release from ballast water**, there are snapshot surveys of particular size classes of organisms collected from a subset of tanks on a subset of ships arriving in selected locations over brief and recent time periods, identified to the lowest taxonomic level possible which nevertheless is often well above that of species (see Table 4-2). As discussed in Chapter 1, organism identities and densities vary within ballast tanks, vessels, and routes (Lavoie et al., 1999; Smith et al., 1999; Wonham et al., 2001; Verling et al., 2005; Lawrence and Cordell, 2010). As a result, the best empirical estimates of organism release from ballast water are recent, local, taxonomically variable subsamples of the process. Again, the degree of uncertainty is not well characterized.

Thus, two difficulties emerge in parameterizing the multi-species risk–release relationship from empirical data. First, neither the dependent nor the independent variable is well resolved. Second, there is a spatial and temporal mismatch between the dependent and independent variables, in that invasion risk is estimated from the outcome of a cumulative century-long historic process, whereas organism release is estimated over a very short time period of months to years (as is evident in Table 4-2). As a result, our ability to rigorously explore the risk–release relationship at the multiple species scale with existing data is greatly limited.

Despite the empirical difficulties, both descriptive and mechanistic modeling frameworks have been developed for the multi-species risk–release relationship, and to some extent parameterized. One response to the absence of robust release data has been to use proxy variables in place of direct measures of propagule pressure. The merits of this strategy are discussed in some detail below, using the examples that follow.

It should be noted that additional theoretical probability and demographic models of species assemblages have developed in the context of island biogeography (MacArthur and Wilson, 1967), local–regional species richness patterns (e.g., Shurin et al., 2000), community assembly (Case, 1990, 1995), and metacommunities (Holyoak et al., 2005). All of these approaches explore the risk–release relationship in its broadest sense. However, since they do not directly address the question of invasion risk vs. organism density, they are unlikely to provide major insight into the question of ballast water standard setting and thus are not reviewed further.

Descriptive Models

As for the single-species scale, statistical models of the multi-species risk–release relationship offer a phenomenological description of a pattern without requiring that the underlying mechanisms be specified. The majority of these studies, recognizing the difficulty of measuring propagule pressure directly, have measured a proxy variable of human activity ranging from population to transport to economic indices.

Some statistical analyses of large-scale invasion vectors have focused on a single transport or economic variable as a substitute for propagule pressure (e.g., Levine and D'Antonio, 2003; Taylor and Irwin, 2004; Ricciardi, 2006; Costello et al., 2007; see examples in Figure 4-8). Others have used multivariable

TABLE 4-2 Spatial and Temporal Scale of Historical Invasion Records, and Spatial, Temporal, and Sampling Scale of Ballasted Organism Surveys in Inland and Coastal Waters of the U.S. and Canada

	Invasion Records		Ballast Surveys		
Location	***Decades***	***Sources***	***Years (N)***	***Sample type (mesh size)***	***Sources***
Laurentian Great Lakes and St. Lawrence Seaway	1840s–2000s	Ricciardi (2006)	1990–91 (86) 2000–02 (39)	water (41 μm, 110 μm) water (unfiltered) sediment	Locke et al. (1993), Subba Rao et al. (1994); Bailey et al. (2005), Duggan et al. (2005)
Chesapeake Bay, MD	1600–2000s	Fofonoff et al. (2009)	1993–94 (60) 1996–97 (7)	water (80 μm) water (80 μm)	Smith et al. (1999) Lavoie et al. (1999)
San Francisco Bay, CA	1850s–1990s	Cohen and Carlton (1998)	—	—	—
Humboldt Bay, CA	1920s–1990s	Wonham and Carlton (2005)	—	—	—
Coos Bay, OR	1940s–1990s	Ruiz et al. (2000), Wonham and Carlton (2005)	1986–91 (159)	water (80 μm)	Carlton and Geller (1993)
Willapa Bay, WA	1930s–1990s	Wonham and Carlton (2005)	—	—	—
Puget Sound, WA	1900s–1990s	Ruiz et al. (2000), Wonham and Carlton (2005)	2001–07 (372)	water (73 μm)	Cordell et al. (2009), Lawrence and Cordell (2010)
Vancouver, BC (and regional waters)	1900s–1990s	Wonham and Carlton (2005)	2000 (15) 2007–08 (23)	water (80 μm) water (unfiltered)	Levings et al. (2004) Klein et al. (2010)
Prince William Sound, AK	1800–1990s	Hines and Ruiz (2000)	1998–1999	water (80 μm)	Hines and Ruiz (2000)

Note: AK, Alaska; BC, British Columbia; CA, California; MD, Maryland; OR, Oregon; WA, Washington. Ballast surveys include studies of N>5 vessels or voyages.

analyses to tease out the relative importance of propagule pressure, again usually by proxy, among other factors contributing to invasion success (ballast water, Drake and Lodge, 2004; plants, Lonsdale, 1999, Dehnen-Schmutz et al., 2007, Castro and Jaksic, 2008, Dawson et al., 2009; earthworms, Cameron and Bayne, 2009; vertebrates, Jeschke and Strayer, 2006; birds, Chiron et al., 2009; fish, Copp et al., 2010) (see Figure 4-8A-C). A variety of linear and non-linear relationships have emerged from these analyses. However, even in the case of a strong statistical relationship, the question of causation must be examined carefully to minimize spurious significant effects caused by confounding variables (Lonsdale, 1999; Figure 4-8D-E) and to distinguish observed patterns from null expectations (Lockwood et al., 2009).

For the case of ballast water, linear risk–release relationships have been estimated in a number of systems (Box 4-3). There are both theoretical assumptions and logistical challenges in developing these models. The first assumption is that total organism number, regardless of the number or abundance of the constituent species, is a reasonable predictor of the number of successfully establishing species. Although these two variables do not have an explicit causal connection, it is intuitively clear that increasing total abundance requires increasing species number, abundance, or both, any of which would be predicted to increase invasion risk. However, the causation is indirect and the precise nature of this relationship is unclear.

The second assumption is that this relationship is linear. Although theory predicts that this relationship should be non-linear, and would be expected to be sigmoidal (as in the single-species case) if Allee effects were operative for species with the highest likelihood of establishment, the trend in a limited dataset may be indistinguishable from linear. Therefore, statistical model fitting should compare multiple candidate models before selecting a linear (or any other) shape.

In fitting such a model to data, the operational challenges quickly become clear. We have estimates of organism density and number of invaders for only a handful of locations (Table 4-2). The density measures have been made with different methods and taxonomic foci, are recent and short-term relative to the accumulation of invaders over decades of ballast water release (Table 4-2), and are patchy and possessed of considerable uncertainty (Verling et al., 2005; Minton et al., 2005; Lawrence and Cordell, 2010). Even if density estimates were entirely accurate and precise, we would not necessarily expect current estimates to predict historical invasion success.

In the face of these difficulties, some authors have used shipping metrics such as vessel number, vessel tonnage, and ballast volume as proxies for propagule pressure (Box 4-3). At first glance, proxy variables appear to offer an appealing way to proceed, since unlike organism density data, vessel traffic data are relatively easy to collect, can be collected retroactively, and might seem to be plausible stand-ins for organism density. However, their use relies on the

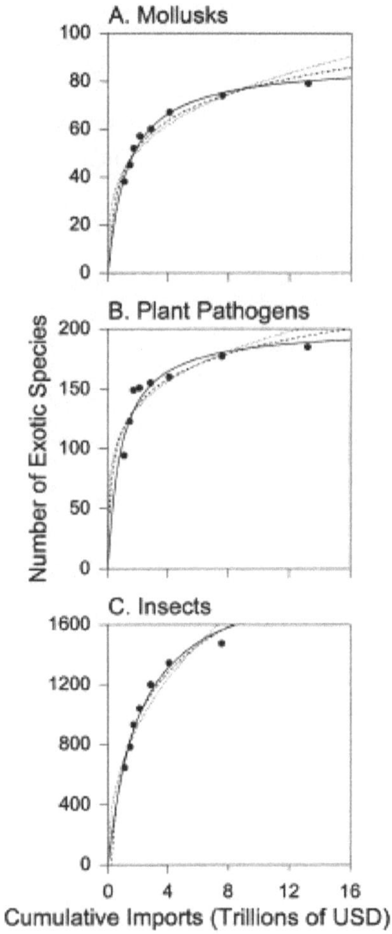

A–C

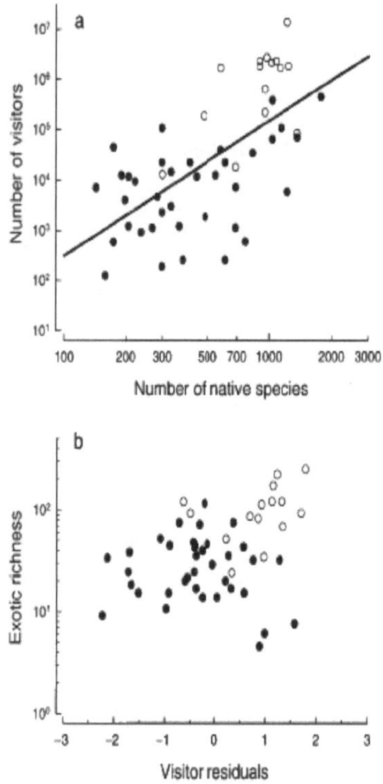

D–E

FIGURE 4-8 Descriptive models of multi-species risk–release relationships. Comparison of log-log (dotted), log-linear (dashed), and Michaelis-Menten (solid) equations fit to number of (A) mollusks, (B) plant pathogens, and (C) insects, vs. cumulative imports over time in the U.S. (Levine and D'Antonio, 2003). (D) Log-log plot of number of visitors vs. number of native plant species in nature reserves worldwide; (E) log-linear plot of number of nonindigenous plant species vs. visitor residuals (non-native plant species as a function of the residuals from the relationship in D) ($p<0.001$) (Lonsdale, 1999). SOURCE: (E) Reprinted, with permission, from Lonsdale (1999). © 1999 by Ecological Society of America.

BOX 4-3
Linear Statistical Multispecies Models

Linear models have been used to estimate the multi-species risk–release relationship for ballast water. To date, the most widely analyzed data at this scale are invasion trends in the Great Lakes. A variety of analyses have been conducted for this system using different data subsets, and different dependent and independent variables. In all the analyses, the data have been parsed into temporal intervals to provide multiple data points for model fitting. The results are not consistent among analyses.

Depending on the data subset, there may or may not appear to be a significant risk–release relationship. Ricciardi (2001) used a linear regression to estimate the rate of all species invasion vs. shipping tonnage in net tons, by decade from 1900 to 1999 (y = 0.062x, r^2 adj = 0.62, p<0.004). This analysis was updated by Ricciardi (2006) for only those free-living invaders assumed to have been introduced by shipping (y = 0.05x; r^2 = 0.516, p<0.019; Figure 4-9A). In contrast, Grigorovich et al. (2003) analyzed Great Lakes invasion data in 5-year intervals from 1959-1999; their data show no clear trend in new invaders as a function of the net tonnage of overseas ballasted traffic, and if anything a negative relationship with the number of overseas ballasted vessels (Figure 4-9B-C).

Linear relationships have been used to estimate a per-ship invasion rate. Drake and Lodge (2004) reanalyzed the data in Ricciardi (2001) against shipping tonnage in metric tons, using a linear regression with a Poisson error distribution (y = 8.47 x 10-8x; p<0.0001). Rescaling by the average ship tonnage, they estimated a per-ship probability of causing an invasion as 0.00044 (95% CI = 0.00008), equivalent to 1 species per 2275 ships or 0.44 invasions per 1000 ships (95% CL 0.36, 0.52). This estimate was based on all nonindigenous species in the Great Lakes, regardless of their presumed vector. In contrast, Costello et al. (2007) used annual data on ship-mediated animal introductions alone from 1959–2000, as a function of number of ships (Figure 4-9D), and obtained a maximum likelihood estimate of 0.14 animal invasions per 1000 ships (95% CL 0.02, 5.2).

Linear relationships have also been used to estimate a per-organism invasion rate for 17 North American ports (Reusser, 2010). These data were not separated into time intervals, so the relationship for each port was based on a single data point. The dependent variable was the total number of established invaders (invertebrates and macroalgae) from 1981–2006 considered to likely to have been introduced by ballast water. The independent variable was the total volume of foreign ballast water discharged from 2005–2007, multiplied on a per-ship basis by a random selection from an empirically determined zooplankton density distribution that spanned eight orders of magnitude (based on

box continues next page

BOX 4-3 Continued

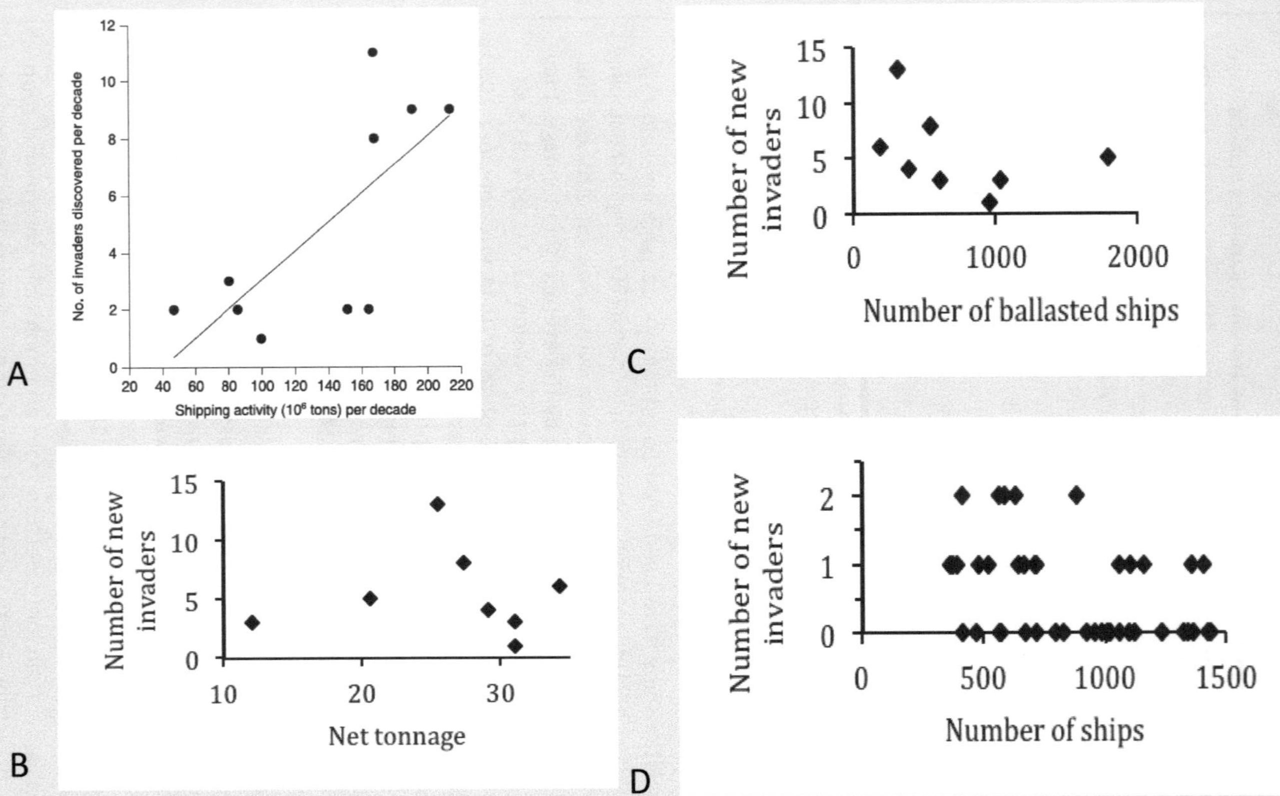

FIGURE 4-9 Invasion patterns in the Great Lakes. (A) The number of invaders scales positively with net shipping tonnage by decade 1900–1999 (Ricciardi, 2006). The same trend is not evident from scatter plots of (B) number of invaders at 5-year intervals from 1959–1999 vs. net tonnage or (C) vs. number of ballasted ships (replotted from Grigorovich et al., 2003) or of (D) annual number of invaders vs. number of ships (plotted from data in Costello et al. 2007, appendices A-B). SOURCE: (A) Reprinted, with permission, from Ricciardi (2006). © 2006 by John Wiley and Sons. (B, C) Reprinted, with permission, from Grigorovich et al. (2003). © 2003 by NRC Research Press. (D) Reprinted, with permission, from Costello et al. 2007. © 2007 by Ecological Society of Amer-

354 ships sampled in four U.S. ports, of which three are included in the 17 analyzed; Minton et al., 2005). Repeated random draws generated a bootstrapped estimate of the median and the first and third quantile invasion rates for each port. Together, these per-capita invasion rates spanned four orders of magnitude from 10^{-11} to 10^{-8}, or one invasion for every 10 million to 10 billion organisms (Reusser, 2010). Interestingly, the data provided no evidence of a strong risk–release relationship across ports, based on either number of vessels or ballast water volume (Figure 4-10A-B).

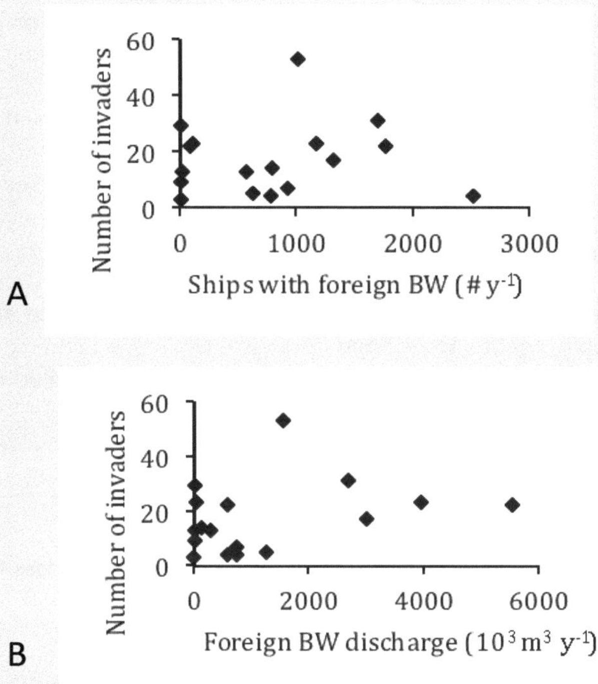

FIGURE 4-10 Across 17 U.S. coastal ports, the number of invaders reported from 1981–2006 shows no strong relationship with (A) number of ships with foreign ballast 2005–2007 or (B) volume of ballast water discharged 2005–2007 (plotted from data in Reusser, 2010, Table 3-2).

box continues next page

BOX 4-3 Continued

The results from these attempts to characterize a multispecies risk–release relationship are ambiguous and highlight the challenges in quantifying propagule pressure. The Great Lakes analyses (Figure 4-9) use shipping traffic variables that in and of themselves do not directly cause invasions, and that have not been tested for their correspondence to organism density. In other words, they are serving as proxies for propagule pressure under the untested assumption that they scale linearly with propagule pressure. The coastal analysis (Reusser, 2010) is an attempt to use a more direct measure of propagule pressure. However, ballast water volume is scaled up assuming the same organism density distribution for all ships, and the relationship is based on a mismatched dataset of invasion data, shipping data, and organism data from different years and ports.

Both Cohen (2005, 2010) and Reusser (2010) have proposed using a linear risk–release relationship to inform ballast water discharge standards [Figure 4-2A; reviewed in Lee et al. (2010)]; Drake and Lodge (2004) used a linear risk–release relationship embedded within a gravity model to investigate risk-reduction strategies. The primary theoretical challenge in developing these approaches is identifying the expected shape of the relationship, particularly given that even total organism number cannot be expected to directly predict species establishment. The primary practical challenge is the current absence of the appropriate data, i.e., spatially and temporally matched variables, and untested or unrepresentative proxy variables.

critical assumption that organism density is homogeneous across tanks and vessels–an assumption that ballast surveys tell us categorically does not hold (Verling et al., 2005; Minton et al., 2005; Lawrence and Cordell, 2010; see also references in Table 4-2). As a result, these variables cannot mechanistically explain the risk–release relationship. Any statistically significant relationship that emerges may represent a spurious correlation. Any non-significant relationship could be the result of a non-representative proxy, or from the absence of a fundamental underlying relationship between risk and release density resulting from the myriad other factors that influence success (see Chapter 3). The results from analyses to date are ambiguous and highlight that proxy variables may not always be reliable predictors of invasion risk, particularly across regions (see Box 4-3). The principle of using proxy variables is not without merit, but it is essential to select and test candidate variables with care before assigning any meaning to their relationships, or lack thereof.

Mechanistic Models

Mechanistic modeling frameworks can be scaled up from single species models to represent the release of multiple species from multiple ships at multiple locations and multiple times. As the mathematical framework of such a model expands, so too do the data requirements for model parameterization and validation.

It is crucial to recognize that it is not possible, mechanistically, to predict the invasion risk associated with the release of an unknown number of unidentified species at unknown abundance, density, and frequency. Any mechanistic multi-species model is necessarily parameterized for a specific group of taxa, and its output is therefore as case-specific as that of a single-species model. To parameterize such a model requires knowing the identities and numbers of all released species, and knowing which of those incipient introductions established and failed. These data can be obtained from controlled experimental studies (e.g., Tilman, 1997; Shurin, 2000; Lee and Bruno, 2009), but not at the full scale of ballast water discharge.

At present, therefore, there are not sufficient taxonomic information or empirical data to parameterize either a probabilistic or a demographic model for all the species in a ballast assemblage. Even if such data and information were available, the time scale mismatch between ballast water discharge and invasion record datasets would still prevent the validation (testing) of a mechanistic model against the empirical data. Nevertheless, one can examine what the framework of such a model might look like with an eye to parameterizing it in the future.

Probabilistic Models

Probabilistic models of the single-species invasion process can be scaled up to create a framework representing the introduction of multiple species, as well as multiple vessels, locations, and releases. This model expresses invasion risk as either the expected number of established species, or the probability that at least one species will establish, in a given time frame. Organism release is specified as the number of individuals of each species released, and may also include separate releases from multiple vessels at multiple locations on multiple occasions (Costello et al., 2007; USCG, 2008). As for single-species models, when the constituent probabilities of the multispecies model are drawn from distributions, the approach is described as a hierarchical probability model (HPM). HPM has the advantage of explicitly representing the known uncertainty in the inherently stochastic invasion process, in the same way that a dynamic demographic model can be made stochastic by drawing parameter values from a distribution.

The multiple-species, multiple-invasion framework readily allows new information about the characteristics of release and establishment (e.g., vessel type and source location; habitat and seasonal differences) to be incorporated into the hierarchical framework. This is particularly useful for evaluating which of many possible scenarios resulting in the observed number of species invasions is best supported by the data. Box 4-4 outlines a general framework for a multi-species HPM of invasion establishment. While this mechanistic hierarchical probability model poses an interesting framework for thinking formally about multi-species and multi-variable risk-release relationships, it has not yet been parameterized or validated with an empirical dataset, and the prospects of doing so are currently remote. Nonetheless, HPM for both single and multi-species scenarios holds advantages over other models because (1) it offers a mechanistic representation of the invasion process, (2) in the absence of detailed distribution data for all the parameters, it can be used in a simplified (i.e., non-hierarchical, point-estimate) version, and (3) as more data become available, it can be easily expanded to incorporate different species, locations, seasons, vessels, etc. See the conclusions of this chapter for a summary recommendation about this approach.

In a somewhat different approach, Costello et al. (2007) adapted a probabilistic model of species introduction and detection over time (Solow and Costello, 2004) to test the relationship between invasion rate and number of vessels arriving annually in the Great Lakes. This analysis highlights the influence of detection lag in confounding our ability to assess the effectiveness of changes in ballast management. Although this relationship is formulated as a probabilistic model, it is based on a proxy variable (number of vessels) so its mechanistic interpretation is unclear. It should be noted that at present, most available data are surrogates of propagule pressure (e.g., number of ships, ballast volume, or ballast discharge) and the number of invaders observed within a given time frame. The existing multispecies models (and most single species models) assume that a reduction in ballast reduces the number of invaders linearly in a system or probabilistically per ship (e.g., Drake and Lodge 2004; Costello, 2007). This may be reasonable given the limited data currently available to construct these models.

Dynamic Demographic Models

Like probabilistic models, dynamic demographic models could in principle be scaled up from the single-species scenario to model the combined risk of many species establishing. Again, such an exercise would require constructing and parameterizing a model with the identity, initial number, and invasion success of each population, and again, the resulting relationship would apply only to that suite of species.

BOX 4-4
Outline of a Simple Multispecies Probabilistic Model

The single-species probabilistic model developed in Box 4-1 is readily extended to a multispecies probabilistic model. Equation 4-3 in Box 4-1 defines the establishment probability for a single species as $p_E = 1\text{-}e^{-\alpha N}$. This equation can be modified to accommodate *S* species, each with its own establishment probabilty p_s. Following the same general approaches described in Shea and Possingham (2000) and USCG (2008), the model then describes S_E, the expected number of species that establish, as

$$S_E = \sum_{s=1}^{S} 1 - e^{-\alpha_i N_i^{c_i}} \quad (4\text{-}8)$$

and p_s, the probability that at least one species establishes, as

$$p_S = 1 - \prod_{i=1}^{S} e^{-\alpha_i N_i^{c_i}} \quad (4\text{-}9)$$

The same principles can be used to extend the model to consider variation in establishment probability across multiple locations (*L*), multiple vessels arriving in those locations (V_L), and so on. In this case, the expected number of species that establish can be written as

$$S_E = \sum_{s=1}^{S} 1 - \prod_{l=1}^{L} \prod_{v=1}^{V_L} e^{-\alpha_{s,l,v} N_{s,l,v}^{c_{s,l,v}}} \quad (4\text{-}10)$$

and the probability that at least one species establishes as

$$p_S = 1 - \prod_{s=1}^{S} \prod_{l=1}^{L} \prod_{v=1}^{V_L} e^{-\alpha_{s,l,v} N_{s,l,v}^{c_{s,l,v}}} \quad (4\text{-}11)$$

box continues next page

BOX 4-4 Continued

These models follow the same logic as the basic one (4-3), where the shape parameters α and c describe the species establishment probability as a function of the initial number of individuals N; the subscripts s, l, and v allow variation among species, locations, and vessels; and the complement of all the propagules failing to establish gives the final probability of establishment. This model does not account for variation in parameter values over time or for potential interactions among species. Nevertheless, parameterizing such a model, particularly in a hierarchical structure where each parameter is characterized as a frequency distribution, would require a tremendous amount of data.

Qualitatively, the overall establishment probability p_s obtained from a multi-species model can only be the same as, or greater than, the largest establishment probability p_E of the constituent species.

To the Committee's knowledge, this approach has not yet been applied to predicting the success of multiple nonindigenous species. However, a related approach using a reaction-diffusion model, which is a standard spatial extension of a demographic model, offers interesting insights. Reaction-Diffusion (R-D) models represent a class of models that were originally developed to model the spread of organisms in continuous time and space (Skellam, 1951). These models were later developed to model the spread of invading organisms across a one- or two-dimensional landscape with the goal of defining the rate of spread as a travelling wave and so provide a description of the rate of spread and the area occupied by the invasion (Okubo et al., 1989). The classical version of these models typically involves several restrictive assumptions including spatial homogeneity and random movement (at least at the population level). However, they are comparatively easy to parameterize, requiring only estimates of per capita rate of population increase and the mean squared displacement per unit time of individuals in the population. Among the advantages of this approach include the ability to approximate the spread of the invading population as a travelling wave (Okubo et al., 1989). This permits the estimation of the rate of spread as a linear function of time, so that the arrival of an invader at a new site could be reasonably estimated under the model assumptions.

These models typically do not provide an estimate of the rate of establishment, although recent applications have attempted this for multiple species using some simplifying assumptions (Drake et al., 2005). In Drake et al. (2005), the authors combined a generic exponential-growth reaction-diffusion model with an allometric relationship between body size and population growth rate to examine the establishment probability of a variety of aquatic species. Their goal was not to predict establishment probabilities of any particular species, but to predict invasion rates over a range of species. Under a series of assumptions, particularly concerning Allee dynamics, this model estimates the risk–release relationship in terms of the proportion of species of a given body size that establish vs. the volume of water released. From this output, the chance of a single invasion by a size class of organism can be predicted as a function of the number and volume of releases, independent of the number of individual organisms. This proposed approach is specific to a size class, and has not yet been validated, but seems reasonable.

Obtaining a Discharge Standard from Multiple-Species Models

Multi-species models represent an attempt to capture the complexity of wholesale ballast water release. Descriptive and mechanistic models can readily be formulated in conceptual and mathematical terms at this scale. In the current absence of data for parameterizing and validating mechanistic models, descriptive statistical models can be developed. However, these must be interpreted with caution given the uncertainty in the estimates, and disconnect between the scales of the independent and dependent variables. The use of proxy variables introduces a further challenge: although well-fitting models may be obtained, proxy variables (such as ballast volume, shipping tonnage, vessel abundance) must be evaluated for their relationship to the direct variable of interest (discharge density) before ascribing any mechanistic meaning to their relationships.

To summarize, a multi-species approach focuses on the assemblage of species released from ballast water. Because of the associated data requirements, descriptive models are more likely than mechanistic models to yield estimates of the risk–release relationship at this scale. Even so, given the uncertainty and mismatch in both the independent and dependent variables, the applicability of any apparent relationships is questionable. Relative to the single-species approach, a multi-species approach has the advantage of being conceptually more realistic in the context of ballast water release, and the disadvantage of being more complex and more difficult to ground in the relevant empirical data.

CONCLUSIONS AND RECOMMENDATIONS

Models are generally useful in environmental management because they provide a transparent framework, force an explicit statement of assumptions, allow us to predict and compare future projections under different management scenarios, and can be updated in their structure and parameter estimates as new information emerges. In principle, a well-supported model of the relationship between invasion risk and organism release could be used to inform a ballast water discharge standard. For a given discharge standard, the corresponding invasion risk could be predicted, or, for a given target invasion risk, the corresponding target release level could be obtained.

Candidate risk–release models developed to date include single- and multiple-species scales, and extend along the spectrum from descriptive to mechanistic in their construction. Mechanistic single-species models require fewer data to parameterize than do mechanistic multi-species models, but do not represent the more realistic scenario of ballast discharge of an assemblage of species. Descriptive single-species models are simpler, but offer none of the predictive advantages of mechanistic ones. Descriptive multi-species models are an appealing tool for investigating large correlative datasets, but are hampered by a current lack of appropriate data.

The rigorous use of models requires that multiple candidate models be formulated and compared in their ability to represent the data. This approach is well established in the population dynamic literature at the single-species scale. However, currently there are insufficient data to distinguish among risk–release relationship models at the multi-species scale. The following conclusions and recommendations identify how models might be put to use at present, and in the future, to help inform a discharge standard.

Ballast water discharge standards should be based on models, and be explicitly expressed in an adaptive framework to allow the models to be updated in the future with new information. Before being applied, it is essential that candidate models be tested and compared, and their compounded uncertainty be explicitly analyzed. Only a handful of quantitative analyses of invasion risk–release relationships thus far have tested multiple models and quantified uncertainty.

The *predicted* shape of the risk–release relationship is non-linear. Inflection points and slope-balance points could provide natural breakpoints for informing a discharge standard. However, the *apparent* shape of the relationship for a given system will depend on the quantity, error, and parameter range of the empirical data, as well as the biology of the species and the nature of the environment.

In the short term, mechanistic *single-species* models are recommended to examine risk–release relationships *for best-case (for invasion) scenario species*. This approach makes sense biologically because in general concerns are only about the small subset of released species that establish as high-impact invaders. Such an approach to setting a standard is conservative and would provide maximum safety against invader establishment.

Candidate best-case-scenario species should be those with life histories that would favor establishment with the smallest inoculum density. Species with the highest probability of establishment relative to inoculum density will have the greatest influence in determining the shape of the risk–release curve. Life history traits promoting such sensitivity to small inoculum density possibly include fast-growth, parthenogenetic or other asexual reproductive abilities, lecithotrophic larvae, etc. Other considerations of best-case species might include those that have a high ecological or economic impact, or are frequently introduced. **The greatest challenge in this approach will be converting the results of small-scale studies to an operational discharge standard.**

Developing a *mechanistic multi-species model* of risk and release, parameterized for an assemblage of best-case scenario species, would only be recommended over the longer term. This model would allow a detailed theoretical investigation of the relationship between total organism number and invasion risk, by permitting the analysis of the risk associated with different species richness and frequency distributions that sum to the same total organism number. The challenges in this approach include the massive time and effort needed to gather the necessary data as well as converting model results to a fleet-wide discharge standard. The implications of these models would therefore be highly specific, no more (and possibly less) informative than those of single-species models, and will require more data and computational effort to construct, parameterize, and validate.

Developing a robust *statistical model* of the risk–release relationship is recommended. It is unclear whether the current lack of a clear pattern across ports reflects a true absence of pattern, or the absence of appropriate data to test this model. Nevertheless, given spatial variation in shipping patterns and environmental variables, it is anticipated that this approach will be more fruitful at a local scale than a nation-wide scale.

Within a region, this relationship should be estimated across multiple time intervals, rather than from a single point. The effect of temporal bin sizes on the shape of the relationship must be examined. The choice of independent variable must be carefully considered. Since long-term historical data on ballast-organism density do not exist, the committee recommends an extremely careful analysis and validation of any proxy variables. The greatest challenge in this approach is the currently insufficient scope and scale of the data. **There is no evidence that any proxy variable used thus far is a reliable stand-in for organism density.**

Finally, models of any kind are only as informative as their input data. In the case of ballast water, both invasion risk and organism density discharged from ballast water are characterized by considerable and largely unquantified, uncertainty. At the multi-species scale in particular, the existing data (historical invasion records vs. recent ballast surveys) are substantially mismatched in time, and patchy in time, space, and taxonomy; current statistical relationships with these or proxy variables are of dubious value. The judicious use of an appropriate model combined with robust data may help inform standard setting in the future.

REFERENCES

Allee, W. 1931. Animal Aggregations. A Study in General Sociology. Chicago: University of Chicago Press.

Andersen, M. 2005. Potential applications of population viability analysis to risk assessments for invasive species. Human and Ecological Risk Assessment 11:1083–1095.

Bailey, S., L. Velez-Espino, O. Johannsson, M. Koops, and C. Wiley. 2009. Estimating establishment probabilities of Cladocera introduced at low density: an evaluation of the proposed ballast water discharge standards. Canadian Journal of Fisheries and Aquatic Sciences 66:261–276.

Barry, K. L., and C. D. Levings. 2002. Feasibility of using the RAMAS metapopulation model to assess the risk of a non-indigenous copepod *(Pseudodiaptomus marinus)* establishing in Vancouver Harbour from ballast water. Canadian Technical Reports Fisheries Aquatic Science 2401.

Bartell, S., and S. Nair. 2004. The establishment of invasive species: an interface between risk analysis and theoretical population ecology. Risk Analysis 20:833–846.

Bertolino, S. 2009. Animal trade and non-indigenous species introduction: the world-wide spread of squirrels. Diversity and Distributions 15:701–708.

Cameron, E., and E. Bayne. 2009. Road age and its importance in earthworm invasion of northern boreal forests. Journal of Applied Ecology 46:28–36.

Carlton, J., and J. Geller. 1993. Ecological roulette: The global transport of nonindigenous marine organisms. Science 261:78–82.

Case, T. 1990. Invasion resistance arises in strongly interacting species-rich model competition communities. Proceedings of the National Academy of Sciences of the United States 87:9610–9614.

Case, T. 1995. Surprising behavior from a familiar model and implications for competition theory. The American Naturalist 146:961–966.

Castro, S. A., and F. M. Jaksic. 2008. How general are global trends in biotic homogenization? Floristic tracking in Chile, South America. Global Ecology and Biogeography 17:524–531.

Caswell, H. 1989. Matrix Population Models: Construction, Analysis and Interpretation. Sunderland, MA: Sinauer Assoc.

Chapman, J., and J. Carlton. 1991. A test of criteria for introduced species: The global invasion by the isopod *Synidotea laevidorsalis* (Miers, 1881). Journal of Crustacean Biology 11:386–400.

Chiron, F., S. Shirley, and S. Kark. 2009. Human-related processes drive the richness of exotic birds in Europe. Proceedings of the Royal Society of London Series B 276:47–53.

Cohen, A. 2005. Memo on a natural invasion rate standard. California State Lands Commission.

Cohen, A. N. 2010. The natural invasion rate model and California's ballast water discharge standards. Presentation to the National Academy of Science/National Research Council Committee on Assessing Numeric Limits for Living Organisms in Ballast Water Washington, DC. June 2, 2010.

Cohen, A. N., and J. T. Carlton. 1998. Accelerating invasion rate in a highly invaded estuary. Science 279(5350):555–558.

Copp, G. H., M. Templeton, and R. E. Gozlan. 2007. Propagule pressure and the invasion risks of non-native freshwater fishes: a case study in England. Journal of Fish Biology 71:148–159, Suppl. D.

Copp, G., L. Vilizzi, and R. Gozlan. 2010. The demography of introduction pathways, propagule pressure and occurrences of non-native freshwater fish in England. Aquatic Conservation: Marine and Freshwater Ecosystems 20:595–601.

Cordell, J., D. Lawrence, N. Ferm, L. Tear, S. Smith, and R. Herwig. 2009. Factors influencing densities of non-indigenous species in the ballast water of ships arriving at ports in Puget Sound, Washington, United States. Aquatic Conservation: Marine and Freshwater Ecosystems 19:322–343.

Costello, C., and A. Solow. 2003. On the pattern of discovery of introduced species. Proceedings of the National Academy of Sciences 100:3321–3323.

Courchamp, F., L. Berec, and J. Gascoigne. 2009. Allee Effects. In: Ecology and Conservation. Oxford: Oxford University Press.

Dawson, W., D. Burslem, and P. Hulme. 2009. Factors explaining alien plant invasion success in a tropical ecosystem differ at each stage of invasion. Journal of Ecology 97:657–665.

Dehnen-Schmutz, K., J. Touza, C. Perrings, and M. Williamson. 2007. The horticultural trade and ornamental plant invasions in Britain. Conservation Biology 21:224–231.

Dennis, B. 2002. Allee effects in stochastic populations. Oikos 96:389–401.

Dennis, B., and S. Lele. 2009. Bayesian methods for hierarchical models: are ecologists making a Faustian bargain? Ecological Applications 19:581–584.

Drake, J. 2004. Allee effects and the risk of biological invasion. Risk Analysis 24:795–802.

Drake, J., K. Drury, D. Lodge, A. Blukacz, and N. Yan. 2006. Demographic stochasticity, environmental variability, and windows of invasion risk for *Bythotrephes longimanus* in North America. Biological Invasions 8:843–861.

Drake, J., and C. Jerde. 2009. Stochastic models of propagule pressure and establishment. Pp. 83–102 In: Bioeconomics of Invasive Species. R. Keller, D. Lodge, M. Lewis, and J. Shogren, editors. Oxford: Oxford University Press.

Drake, J. M., and D. M. Lodge. 2004. Global hot spots of biological invasions: evaluating options for ballast-water management. Proceedings of the Royal Society of London Series B-Biological Sciences 271(1539):575–580.

Drake, J., D. Lodge, and M. Lewis. 2005. Theory and preliminary analysis of species invasions from ballast water: controlling discharge volume and location. American Midland Naturalist 154:459–470.

Duggan, I., C. van Overdijk, S. Bailey, P. Jenkins, H. Limén, and H. MacIsaac. 2005. Invertebrates associated with residual ballast water and sediments of cargo-carrying ships entering the Great Lakes. Canadian Journal of Fisheries and Aquatic Sciences 62:2463–2474.

Dullinger, S., I. Kleinbauer, J. Peterseil, M. Smolik, and F. Essl. 2009. Niche based distribution modelling of an invasive alien plant: effects of population status, propagule pressure and invasion history. Biological Invasions 11:2401–2414.

Dunstan, P., and N. Bax. 2007. How far can marine species go? Influence of population biology and larval movement on future range limits. Marine Ecology Progress Series 344:15–28.

Fitzpatrick, M., E. Preisser, A. Ellison, and J. Elkinton. 2009. Observer bias and the detection of low-density populations. Ecological Applications 19:1673–1679.

Fofonoff, P. W., G. M. Ruiz, A. H. Hines, B. D. Steves, and J. T. Carlton. 2009. Four Centuries of Estuarine Biological Invasions in the Chesapeake Bay Region. Pp. 479–506 In: Biological Invasions in Marine Ecosystems: Ecological, Management, and Geographic Perspectives. Rilov, G. and J. Crooks (editors). Berlin: Springer-Verlag.

Gotelli, N. J. 1991. Metapopulation Models: The Rescue Effect, the Propagule Rain, and the Core-Satellite Hypothesis. American Naturalist 138(3):768–776.

Grigorovich, I. A., R. I. Colautti, E. L. Mills, K. Holeck, and H. J. MacIsaac. 2003. Ballast-mediated animal introductions in the Laurentian Great Lakes: retrospective and prospective analysis. Canadian Journal of Fisheries and Aquatic Sciences 60:740–756.

Haccou, P., and Y. Iwasa. 1996. Establishment probability in fluctuating environments: a branching process model. Theoretical Population Biology 50:254–280.

Hastings, A., K. Cuddington, K. Davies, C. Dugaw, S. Elmendorf, A. Freestone, S. Harrison, M. Holland, J. Lambrinos, U. Malvadkar, B. Melbourne, K. Moore, C. Taylor, and D. Thomson. 2005. The spatial spread of invasions: new developments in theory and evidence. Ecology Letters 8:91–101.

Herborg, L.-M., C. Jerde, D. Lodge, G. Ruiz, and H. MacIsaac. 2007. Predicting invasion risk using measures of introduction effort and environmental niche models. Ecological Applications 17:663–674.

Herborg, L. M., P. O'Hara, and T. W. Therriault. 2009. Forecasting the potential distribution of the invasive tunicate *Didemnum vexillum*. Journal of Applied Ecology 46(1):64–72.

Hines, A. H., and G. M. Ruiz. 2000. Biological invasions of cold-water coastal ecosystems: Ballast-mediated introductions in Port Valdez / Prince William Sound, Alaska. Final Report, Regional Citizens Advisory Council of Prince William Sound.

Holyoak, M., M. Leibold, and R. Holt. 2005. Metacommunities: Spatial Dynamics and Ecological Communities. Chicago: University of Chicago Press.

Hulme, P. 2009. Trade, transport and trouble: managing invasive species pathways in an era of globalization. Journal of Applied Ecology 46:10–18.

Jerde, C., C. Bampfylde, and M. Lewis. 2009. Chance establishment for sexual, semelparous species: Overcoming the Allee effect. American Naturalist 173:734–746.

Jerde, C., and J. Bossenbroek. 2009. Uncertain invasions: a biological perspective. In: Bioeconomics of Invasive Species. R. Keller, D. Lodge, M. Lewis, and J. Shogren, editors. Oxford: Oxford University Press.

Jerde, C., and M. Lewis. 2007. Waiting for invasions: A framework for the arrival of nonindigenous species. American Naturalist 170:1–9.

Jeschke, J., and D. Strayer. 2006. Determinants of vertebrate invasion success in Europe and North America. Global Change Biology 12:1608–1619.

Jongejans, E., O. Skarpaas, P. Tipping, and K. Shea. 2007. Establishment and spread of founding populations of an invasive thistle: The role of competition and seed limitation. Biological Invasions 9:317–325.

Klein, G., K. MacIntosh, I. Kaczmarska, and J. Ehrman. 2010. Diatom survivorship in ballast water during trans-Pacific crossings. Biological Invasions 12:1031–1044.

Kramer, A., and J. Drake. 2010. Experimental demonstration of population extinction due to a predator-driven Allee effect. Journal of Animal Ecology 79:633–639.

Lavoie, D., L. Smith, and G. Ruiz. 1999. The potential for intracoastal transfer of non-indigenous species in the ballast water of ships. Estuarine Coastal and Shelf Science 48:551–564.

Lawrence, D., and J. Cordell. 2010. Relative contributions of domestic and foreign sourced ballast water to propagule pressure in Puget Sound, Washington, USA. Biological Conservation 143:700–709.

Lee II, H., D. Reusser, M. Frazier, and G. Ruiz. 2010. Density Matters: Review of Approaches to Setting Organism-Based Ballast Water Discharge Standards.

EPA/600/R-10/031. EPA Office of Research and Development, National Health and Environmental Effects Research Laboratory, Western Ecology Division.

Lee, S., and J. Bruno. 2009. Propagule supply controls grazer community structure and primary production in a benthic marine ecosystem. Proceedings of the National Academy of Sciences of the United States 106:7052–7057.

Leung, B., J. Drake, and D. Lodge. 2004. Predicting invasions: propagule pressure and the gravity of Allee effects. Ecology 85:1651–1660.

Leung, B., and N. Mandrak. 2007. The risk of establishment of aquatic invasive species: joining invasibility and propagule pressure. Proceedings of the Royal Society of London Series B 274:2603–2609.

Levine, J. M., and C. M. D'Antonio. 2003. Forecasting biological invasions with increasing international trade. Conservation Biology 17(1):322–326.

Levings, C., J. Cordell, S. Ong, and G. Piercey. 2004. The origin and identity of invertebrate organisms being transported to Canada's Pacific coast by ballast water. Canadian Journal of Fisheries and Aquatic Sciences 61:1–11.

Lewis, M., M. Neubert, H. Caswell, J. Clark, and K. Shea. 2005. A guide to calculating discrete time invasion rates from data. In: Conceptual Ecology and Invasions Biology: Reciprocal Approaches to Nature. M. Cadotte, S. McMahon, and T. Fukami, editors. New York: Springer.

Locke, A., D. Reid, H. Van Leeuwen, W. Sprules, and J. Carlton. 1993. Ballast water exchange as a means of controlling dispersal of fresh-water organisms by ships. Canadian Journal of Fisheries and Aquatic Sciences 50:2086–2093.

Lockwood, J., P. Cassey, and T. Blackburn. 2009. The more you introduce the more you get: the role of colonization pressure and propagule pressure in invasion ecology. Diversity and Distributions 15:904–910.

Lonsdale, W. 1999. Global patterns of plant invasions and the concept of invasibility. Ecology 80:1522–1536.

Lubina, J., and S. Levin. 1988. The spread of reinvading species: range expansion in the California sea otter. American Naturalist 131:526–543.

Lutscher, F., E. McCauley, and M. Lewis. 2007. Spatial patterns and coexistence mechanisms in systems with unidirectional flow. Theoretical Population Biology 71:267–277.

Lutscher, F., R. Nisbet, and E. Pachepsky. 2010. Population persistence in the face of advection. Theoretical Ecology 3:271–284.

MacArthur, R., and E. Wilson. 1967. The Theory of Island Biogeography. Monographs in Population Biology. Princeton, NJ: Princeton University Press.

Memmott, J., P. G. Craze, H. M. Harman, P. Syrett, and S. V. Fowler. 2005. The effect of propagule size on the invasion of an alien insect. Journal of Animal Ecology 74:50–62.

Minton, M., E. Verling, A. Miller, and G. Ruiz. 2005. Reducing propagule supply and coastal invasions via ships: effects of emerging strategies. Frontiers in Ecology and the Environment 3:304–308.

Morris, W., and D. Doak. 2002. Quantitative Conservation Biology: Theory and Practice of Population Viability Analysis. Sunderland, MA: Sinauer Associates.

Neubert, M., and I. Parker. 2004. Projecting rates of spread for invasive species. Risk Analysis 24:817–831.

Okubo, A., P. Maini, M. Williamson, and J. Murray. 1989. On the spatial spread of the gray squirrel in Britain. Proceedings of the Royal Society of London Series B 238:113–125.

Pachepsky, E., F. Lutscher, R. Nisbet, and M. Lewis. 2005. Persistence, spread and the drift paradox. Theoretical Population Biology 67:61–73.

Peterson, A., and D. Vieglais. 2001. Predicting species invasions using ecological niche modeling: new approaches from bioinformatics attack a pressing problem. BioScience 51:363–371.

Parker, I. 2000. Invasion dynamics of *Cytisus scoparius*: A matrix model approach. Ecological Applications 10(3):726–743.

Ponciano, J. M., M. L. Taper, B. Dennis, and S. R. Lele. 2009. Hierarchical models in ecology: confidence intervals, hypothesis testing, and model selection using data cloning. Ecology 90(2):356–362.

Reusser, D. 2010. Biogeography of Nonindigenous Species: From Description to Prediction. Corvallis, OR: Oregon State University.

Ricciardi A. 2001. Facilitative interactions among aquatic invaders: is an "invasional meltdown" occurring in the Great Lakes? Canadian Journal of Fisheries and Aquatic Sciences 58(12):2513–2525.

Ricciardi, A. 2006. Patterns of invasion in the Laurentian Great Lakes in relation to changes in vector activity. Diversity and Distributions 12:425–433.

Ruiz, G., and J. Carlton. 2003. Invasion vectors: A conceptual framework for management. In: Invasive species: Vectors and management strategies. G. Ruiz and J. Carlton (editors). Washington, DC: Island Press.

Ruiz, G., P. Fofonoff, J. Carlton, M. Wonham, and A. Hines. 2000. Invasion of coastal marine communities in North America: Apparent patterns, processes, and biases. Annual Reviews in Ecology and Systematics 31:481–531.

Shea, K., and H. Possingham. 2000. Optimal release strategies for biological control agents: An application of stochastic dynamic programming to population management. Journal of Applied Ecology 37(1):77–86.

Shigesada, N., and K. Kawasaki. 1997. Biological Invasions: Theory and Practice. Oxford: Oxford University Press.

Shurin, J. 2000. Dispersal limitation, invasion resistance, and the structure of pond zooplankton communities. Ecology 81:3074–3066.

Shurin, J., J. Havel, M. Leibold, and B. Pinel-Alloul. 2000. Local and regional zooplankton species richness: a scale-independent test for saturation. Ecology 81:3062–3073.

Skellam, J. 1951. Random dispersal in theoretical populations. Biometrika 38:196–218.

Smith, L., M. Wonham, L. McCann, G. Ruiz, A. Hines, and J. Carlton. 1999. Invasion pressure to a ballast-flooded estuary and an assessment of inoculant survival. Biological Invasions 1:67–87.

Solow, A., and C. Costello. 2004. Estimating the rate of species introductions from the discovery record. Ecology 85:1822–1825.

Suarez, A., D. Holway, and P. Ward. 2005. The role of opportunity in the unintentional introduction of nonnative ants. Proceedings of the National Academy of Sciences of the United States 102:17032–17035.

Taylor, B., and R. Irwin. 2004. Linking economic activities to the distribution of exotic plants. Proceedings of the National Academy of Sciences of the United States 101:17725–17730.

Tilman, D. 1997. Community invasibility, recruitment limitation, and grassland biodiversity. Ecology 78:81–92.

USCG. 2008. United States Coast Guard Ballast Water Discharge Standards Draft Programmatic Environmental Impact Statement.

Verling, E., G. M. Ruiz, L. D. Smith, et al. 2005. Supply-side invasion ecology: characterizing propagule pressure in coastal ecosystems. Proceedings of the Royal Society B-Biological Sciences 272(1569):1249–1256.

Wonham, M., and J. Carlton. 2005. Trends in marine biological invasions at local and regional scales: the Northeast Pacific Ocean as a model system. Biological Invasions 7:369–392.

Wonham, M., and M. Lewis. 2009. Modeling marine invasions: current and future approaches. Pp. 71–105 In: Marine Bioinvasions: Ecology, Conservation, and Management Perspectives. G. Rilov and J. Crooks, editors. Berlin: Springer-Verlag.

Wonham, M., and E. Pachepsky. 2005. A null model of temporal trends in biological invasion records. Ecology Letters 9:663–672.

Wonham, M. J., W. C. Walton, G. M. Ruiz, A. M. Frese, and B. S. Galil. 2001. Going to the source: role of the invasion pathway in determining potential invaders. Marine Ecology-Progress Series 215:1–12.

Wood, D., J. Koprowski, and P. Lurz. 2007. Tree squirrel introduction: A theoretical approach with population viability analysis. Journal of Mammalogy 88(5):1271–1279.

5

Other Approaches to Setting a Ballast Water Discharge Standard

The EPA and USCG seek a scientifically based process for determining a ballast water discharge standard that will protect against the establishment of new nonindigenous aquatic species in the nation's waterbodies and protect and preserve resident species and other beneficial uses in these systems. In the absence of data and models necessary to support a science-based quantitative approach (see Chapter 4), expert opinion has been a common alternative to modeling to advance discharge standards.

Expert opinion can be defined as a process by which a person, or team of individuals, with a high degree of knowledge pertaining to a particular topic draws upon that knowledge to propose an estimate. In the area of standards for living organisms in ballast water discharge, expert opinion processes have taken a number of forms and, as observed by Lee et al. (2010), produced a wide range of outcomes. This chapter discusses strengths and weaknesses of the expert-opinion approach and reviews three expert opinion-based standards for living organisms in ballast discharge: the (1) IMO standard setting approach, (2) California's Zero Detectable Living Organism standard setting approach, and (3) Natural Invasion Rates as a basis for a standard. The Natural Invasion Rate method put forward by Cohen (2005, 2010) starts with a desired invasion rate based on expert opinion, and then combines it with a quantitative risk–release relationship (much like those discussed in Chapter 4), to derive allowable discharge levels. This chapter considers only the expert opinion-based portion of the method.

EXPERT OPINION AS AN APPROACH TO DECISION-MAKING

Expert opinion has been applied in the field of biological invasions in a number of ways. First, expert opinion has been used to identify nonindigenous species that may pose problems if introduced to a new country (see IUCN, 2000; European Environment Agency, 2010; and Kolar and Lodge, 2002). The Canadian government developed a risk assessment for silver and bighead carp species based upon expert opinion, which resulted in a regulation banning live sales by fish retailers (Mandrak and Cudmore, 2004). Expert opinion also has been used to screen for invasive, weedy plants and insect pests, among others (see Tucker and Richardson, 1995; Daehler et al., 2004; Paini et al., 2010).

An inherent advantage of the expert opinion approach is that it can produce estimates in the absence of robust data sets. Some models based on expert opinion can be cross-checked using other methodologies, allowing researchers and managers to consider the biological realism of their predictions (Phelong et al., 1999; Paini et al., 2010). Nonetheless, there often remains considerable uncertainty in the output from expert opinion, especially when there are critical gaps in understanding underlying mechanisms.

Expert opinion can be greatly affected by problem framing, context dependence, and motivational bias by participants (Burgman, 2005). Thus, a key problem is that very different outcomes may result from different parties considering the same problem. It is also difficult to test the output for validity since critical considerations are often implicit in the process. Consensus-based expert opinion by cross-disciplinary teams may help eliminate blind spots in the analysis in terms of types of expertise, but these exercises often must limit the number of experts per discipline to one individual. As a result, inherent biases within an expert become a greater concern.

Maguire (2004) concluded that risk assessment based upon expert opinion is often conflated with risk management, and that assessments of risk based on individual judgment often incorporate personal outcome preferences. A further problem is the influence of incomplete information on decision making. For example, highly invasive species including the emerald ash borer (*Agrilus planipennis*) and quagga mussel (*Dreissena rostriformis bugensis*) had little or no invasion history prior to colonizing North America, and expert opinion based on invasion history likely would have concluded that these species pose little risk. While risk assessments based upon performance (effect) in native regions may be informative and are commonly conducted, the colonization process can result in evolutionary (genetic) changes, where native and introduced populations differ substantially. Moreover, invasions by definition place species in novel environments, with a different assemblage of organisms. Thus, native populations may be poor predictors of population dynamics, interactions, and effects in an invaded area.

Data on the concentration of living organisms in ballast water and on the probability of invasions occurring in U.S. waterbodies as a result of ballast water are extremely limited. Thus, despite its drawbacks, expert opinion has understandably played a major role in ballast discharge standard-setting to date, and it will likely be involved as part of the risk management decision-making process into the future.

IMO STANDARD SETTING APPROACH

The best-known example of an expert opinion-based standard setting process for ballast water is the agreement reached at the 2004 Diplomatic Conference of the International Maritime Organization (IMO), which included 74 IMO member countries, the European Community, and 18 non-governmental organizations (Gollasch, 2005). This agreement was formalized as the IMO *International Convention for the Control and Management of Ships' Ballast Water and Sediments* and was signed by 67 IMO member countries. As of May 2011, the Convention has been ratified by 28 countries collectively accounting for over 28 percent of global merchant shipping (IMO, 2011). The IMO Convention will come into force one year after not less than 30 States representing 35 percent of the world's merchant tonnage have ratified without reservation.

In this rendition, the expert-opinion process involved at least two sets of experts (there was some overlap) and two sets of rules. Specifically, the IMO Marine Environment Protection Committee (MEPC) requested the assistance of a Study Group (on Ballast Water and Other Ship Vectors) of the International Convention for the Exploration of the Sea (ICES), and also a group from the shipping industry to propose a starting point for IMO negotiations to achieve discharge standards. The ICES Study Group, predominantly aquatic invasion biologists, reviewed available data sets to date related to concentrations of organisms in untreated ballast, characterizing the range and distribution across many vessels for different types of organisms. This expert group recommended that to "significantly reduce the risk of invasions associated with ballast water beyond the present situation, permissible discharge concentrations identified by any treatment/management standards should fall greatly below the median values observed presently for untreated/unmanaged ballast water." This would have the effect of reducing high concentration discharges, and the group considered empirical and theoretical evidence indicating that invasion likelihood increases with increasing concentration (International Council for the Exploration of the Sea, 2003).

Recognizing that biological invasions *via* ballast water occur across a broad span of phyla and size classes, and that organisms comprising these groups differ tremendously in abundance in both nature and ballast water, the ICES study group recommended that the IMO performance standard classify biota contained in ballast water into categories based on size or taxonomic group. This group made recommendations of discharge standards for two of these sizes classes: greater than 50 μm (zooplankton) and between ten and 50 μm (phytoplankton) to achieve concentrations three orders of magnitude below observed median values in untreated ballast (the recommended standards correspond to 0.4 zooplankton per m^3 and 13.3 phytoplankton per liter). Because the recommended approach considers total abundance of organisms in a particular size class (i.e., a guild) in a ship's discharge, the risk associated with release of any one species would be expected to be lower since ballast water communities usually comprise numerous species.

The first stage of the process for crafting the IMO standard represented the consensus expert opinion of the ICES Study Group experts involved. The output of this analysis was then transmitted to the relevant IMO committees and working groups to be considered along with operational concerns cited by industry. The standard that emerged from this two stage process places numerical limits on abundances in treated ballast effluent of less than ten viable organisms per m^3 greater than or equal to 50 μm in minimum dimension (zooplankton and zoobenthos) and less than ten viable organisms per ml less than 50 μm but greater than or equal to 10 μm in minimum dimension (including phytoplankton and protists). The IMO also incorporated into the standard toxigenic *Vibrio cholerae* (serotypes O1 and O139), the etiologic agents of pandemic cholera, and two routinely used fecal-indicator bacteria, *E. coli* and intestinal Enterococci. These latter two are frequently used to assess the safety of recreational surface waters, e.g., swimming beaches. Some are concerned that the allowable levels of live organisms in the standard generally are too high to be protective. In particular, the USCG during negotiations for the IMO agreement indicated that standards needed to be considerably more stringent (i.e., 100 times) to be protective, approaching more closely recommendations from the ICES Study Group. Regardless, it appears that the IMO D-2 standard represents a significant reduction in concentrations beyond ballast water exchange, especially for the largest size range (International Council for the Exploration of the Sea 2003; Minton et al., 2005).

ZERO-DETECTABLE DISCHARGE STANDARD

The State of California's existing and anticipated standard for larger planktonic organisms in discharged ballast water is "zero detectable" living or culturable organisms. Like the IMO standard setting process, the process associated with the genesis of this standard also involved the opinions of technical experts (biologists) as well as industry and other groups (California State Lands Commission, 2010). The expert group recognized that (a) the probability of invasion declined with concentration and (b) there was great uncertainty surrounding the risk–release relationship. This group took a precautionary approach, while acknowledging that enforcement against a standard of zero living organisms has practical constraints. Unlike the IMO D-2 standard, the zero detectable discharge standard applies to all size classes and taxa (if the word "living" may be considered applicable to viruses) and is to be put in force by January 2020 (Faulkner, 2010). [The interim standard is zero detectable living organisms for taxa >50μm, 0.01 organisms per ml for sizes between 10 and 50 μm, and no more than 10^3 bacteria and no more than 10^4 viruses per 100 ml of ballast water discharge. Interim standards are to be in place between 2009 and 2016 (Faulkner, 2010).]

This approach to a standard is appealing in its straightforwardness, but as Lee et al. (2010) argue, its rationale is predicated on an actual zero discharge, while its practical application can differ substantially from zero *detectable* discharge. The zero-detectable concept is constrained heavily by the sampling regime employed and associated sensitivity, which will depend on the analysts' technical capabilities and financial resources. Recent advances in molecular ecology, such as use of species' environmental DNA (e.g., Mahon et al., 2011; Jerde et al., 2011), are expanding our capability to detect rare organisms in water, although none can yet unequivocally decipher whether all organisms detected are living or dead.

Because of the practical constraints in detecting living organisms when they are rare, the zero-detectable standard may be operationally no more stringent than other standards. The lower the density of living organisms achieved in the effluent, the more difficult the task of detecting organisms and determining whether the water is compliant with the zero-detectable standard. Lee et al. (2010) nicely summarized this by estimating the uncertainty inherent in sampling different volumes of water (their Table 18). For example, even should a 10 m^3 sample (a huge sample) of ballast water contain no organisms, the actual concentration of organisms in the tank could be as high as 0.3 per m^3. Arguably, there could be fine tuning of the Lee et al. (2010) analysis should the distribution of organisms be found to be other than Poisson, but any such change would only alter the levels of uncertainly associated with a particular sampling regime. More recently, Miller et al. (2011) also considered organisms with a Poisson distribution in ballast water effluent. They demonstrated that the density of organisms and volume sampled had strong effects on decisions to reject whether ballast water is compliant with a selected standard, while the permissible type I error rate (i.e., false positives or rejection of compliant ballast) had slightly less influence as modeled.

Thus, while a laudable ideal, the zero-detectable standard is functionally defined by the ability to characterize concentrations of organisms at low densities. If organisms are detected, it is clear a ship's ballast effluent is in violation. However, critically, the absence of live organisms in a sample or set of samples does not provide sufficient information to accurately assess densities, and there is always a non-zero probability that organisms are present below some threshold (defined by the specific sampling effort imposed). Thus, while California's standards clearly represent a reduction in concentrations below those observed in ballast water exchange, the exact discharge standard is largely undefined and contingent on sampling protocols, representing an operational definition that is driven by sampling statistics.

NATURAL INVASION RATES

The Natural Invasion Rate (NIR) approach to standard setting incorporates both a risk assessment and a risk management component. The risk assessment component assumes that an empirical, linear relationship can be determined between the rate of organism release in ballast water and the resulting rate of new species invasions (see Chapter 4). The risk management component proposes an acceptable target invasion rate based on expert opinion from which the corresponding target discharge rate is determined. In this case, expert opinion has concluded that the *natural (i.e., non-anthropogenic, background) invasion rate* should be the target ballast-mediated invasion rate. This discussion focuses on this latter component of the approach (i.e., the risk management step).

Cohen (2005) identified some difficulties inherent in estimating natural invasion rates and provided coarse estimates based on expert opinion that span three orders of magnitude. Providing any validation or better assessment is problematic because there is insufficient information to determine a natural invasion rate. The scale of available paleontological data is inconsistent with that needed for application of a NIR standard. For example, natural port-to-port invasion rate data are required, whereas available data refer to large-scale dispersal events (e.g., trans-Pacific regional migrations; see Vermeij, 1991) and, thus, are not an appropriate comparator. It is expected that natural estuary-to-estuary trans-Pacific migrations would be much lower than the measured regional migrations. While some natural invasion rate data are available, in terms of Pleistocene migrations and faunal shifts along the California coast, including ancient embayments (Roy et al., 1995), they do not include trans-Pacific species movements. In addition, even if data were available for one estuary, differences likely exist among different recipient ecosystems and would need to be taken into account.

These problems with estimating the natural invasion rate interfere with the potential to obtain a meaningful organism discharge standard that would be equivalent in its effect for all locations, or that would indeed be a desirable target. There is an assumption within the NIR approach that an acceptable invasion rate is less than the natural invasion rate. Such assumptions make it clear that NIR is based on expert opinion as much as the IMO standards and the zero-detectable standard of CA.

CONCLUSIONS

The data available from which to derive scientifically based standards for living organisms in ballast discharge have been incomplete and insufficient. This situation results partly from the degree of variability possible across ballast discharges, the difficulty in garnering representative samples of live organisms in ballast discharge and the lack of a systematic approach for doing so, and the lack of baseline measures to detect the presence of newly established populations of nonindigenous organisms in U.S. waters. A discharge standard based on expert opinion has provided a starting point for the regulatory process and can facilitate progression to a scientific model. For example, as ships attempt to meet the IMO D-2 standard, much of the variation in potential densities of live organisms in ballast discharge will be modulated. Advancing scientific understanding of the probability of invasion associated with such discharges, and the performance of existing standards to reduce invasion outcomes, requires quantitative measures of both the discharge and also associated invasions. While compliance testing provides a potential framework to achieve part of this goal, the broader integrative approach required has yet to develop (see Chapter 6).

In the area of discharge limits for live organisms in ballast water, expert opinion has been tapped as an alternative to scientifically derived standard setting processes in light of scientific uncertainties. While each expert opinion process discussed above has conceptual merit, each is compromised by assumptions, data limitations, or operational difficulties. Despite these drawbacks, expert opinion has resulted in various standards (e.g., IMO D-2) that could provide a manageable baseline for developing scientific models that can be used to quantitatively determine ballast water discharge standards.

REFERENCES

Blaschko, M. B., G. Holness, M. A. Mattar, D. Lisin, P. E. Utgoff, A. R. Hanson, H. Schultz, E. M. Riseman, M. E. Sieracki, W. M. Balch, and B. Tupper. 2005. Automatic in situ identification of plankton. Proceedings of the Seventh IEEE Workshop on Applications of Computer Vision 79–86. doi 10.1109/ACVMOT.2005.29.

Burgman, M. 2005. Risks and decisions for conservation and environmental management. Cambridge, UK: Cambridge University Press.

California State Lands Commission. 2010. 2010 Assessment of the efficacy, availability, and environmental impacts of ballast water treatment systems for use in California waters. Sacramento, CA: California State Lands Commission.

Cohen, A. N. 2005. Memo on a natural invasion rate standard. Appendix 5 In: California State Lands Commission Report on Performance Standards for Ballast Water Discharges in California Waters. Report produced for California State Legislature. M. Falkner, L. Takata, and S. Gilmore, editors.

Cohen, A. N. 2010. The natural invasion rate model and California's ballast water discharge standards. Presentation to the National Academy of Science/National Research Council Committee on Assessing Numeric Limits for Living Organisms in Ballast Water Washington, DC. June 2, 2010.

Daehler, C. C., J. S. Denslow, S. Aansari, and H.-C. Kuoa. 2004. Risk-assessment system for screening out invasive pest plants from Hawaii and other Pacific Islands. Conservation Biology 18:360–368.

European Environment Agency. 2010. Invasive alien species in Europe – Assessment published May 2010. Available online at: *http://www.eea.europa.eu/data-and-maps/indicators/invasive-alien-species-in-europe/invasive-alien-species-in-europe*. Last accessed June 2, 2011.

Faulkner, M. 2010. Presentation to the National Academy of Science/National Research Council Committee on Assessing Numeric Limits for Living Organisms in Ballast Water. June 4, 2010. Washington, DC.

Gollasch, S. 2005. IMO BWMC: History, present status and challenges to the research society. Presentation to the Ballast water introductions of alien species into the Baltic Sea, Palanga.

International Council for the Exploration of the Sea. 2003. Comments on draft Regulation E-2. Submitted to the Marine Environment Protection Committee, MEPC 49/2/21.

IMO (International Maritime Organization). 2011. Available online at *http://www.imo.org/OurWork/Environment/BallastWaterManagement/Pages/Default.aspx#5.* Last accessed June 2, 2011.

IUCN (International Union for the Conservation of Nature). 2000. 100 of the World's Worst Invasive Alien Species A selection from the Global Invasive Species Database. Available online at: *http://www.issg.org/booklet.pdf.* Last accessed June 2, 2011.

Jerde, C., A. R. Mahon, W. L. Chadderton, and D. M. Lodge. 2011. 'Sight-unseen' detection of rare aquatic species using environmental DNA. Conservation Letters doi: 10.1111/j.1755-263X.2010.00158.

Kolar, C. S., and D. M. Lodge. 2002. Ecological predictions and risk assessment for alien fishes in North America. Science 298:1233–1236.

Lee II, H., D. A. Reusser, M. Frazier, and G. Ruiz. 2010. Density Matters: Review of Approaches to Setting Organism-Based Ballast Water Discharge Standards. EPA/600/R-10/031. EPA Office of Research and Development, National Health and Environmental Effects Research Laboratory, Western Ecology Division.

Maguire, L. A. 2004. What can decision analysis do for invasive species management? Risk 24:859–868.

Mahon, A. R., M. Barnes, S. Senapati, J. Darling, H.-C. Chang, J. L. Feder, and D. M. Lodge. 2011. Molecular detection of invasive species in heterogeneous mixtures using a microfluidic carbon platform. PLoS One 6(2):e17280.

Miller, A. W., M. Frazier, G. E. Smith, E. S. Perry, G. M. Ruiz, and M. N. Tamburri. 2011. Enumerating sparse organisms in ships' ballast water: why counting to 10 is not so easy. Environmental Science and Technology (in press). dx.doi.org/10.1021/es102790d.

Minton, M., E. Verling, A. W. Miller, and G. M. Ruiz. 2005. Reducing propagule supply by ships to limit coastal invasions: effects of emerging strategies. Frontiers in Ecology and the Environment 6:304–308.

Paini, D. R., S. P. Worner, D. C. Cook, P. J. DeBarro, and M. B. Thomas. 2010. Using a self-organizing map to predict invasive species: sensitivity to data errors and a comparison with expert opinion. Journal of Applied Ecology 47:290–298.

Phelong, P. C., P. A. William, and S. R. Halloy. 1999. A weed risk assessment model for use as a biosecurity tool evaluating plant introductions. Journal of Environmental Management 57:239–251.

Roy, K., D. Jablonski, and J. W. Valentine. 1995. Thermally anomalous assemblages revisited: patterns in the extraprovincial range shifts of Pleistocene marine mollusks. Geology 23:1071–1074.

Tucker, K. C., and D. M. Richardson. 1995. An expert system for screening potentially invasive alien plants in South African Fynbos. Journal of Environmental Management 44:309–338.

Vermeij, G. J. 1991. When biotas meet: understanding biotic interchange. Science 253: 1099–1104.

6
The Path Forward

INTRODUCTION

This chapter recommends a path forward to inform decisions on ballast discharge standards to reduce the probability of invasions. This path is bracketed by the regulatory climate, the myriad of variables that affect whether organisms in discharged ballast are likely to lead to an invasion, and the state of the science to evaluate the risk–release relationship associated with ballast water discharge.

It is abundantly clear that reducing propagule pressure will reduce the probability of invasions, when controlling for all other variables. There is both strong theoretical and empirical support for this, across a diverse range of habitats, geographic regions, and types of organisms (Sax et al., 2005; Lockwood et al., 2007; Davis, 2009). The key issue, however, is the ability to characterize quantitatively the risk–release relationship, with the goal of functionally describing the incremental reduction in invasion probability achieved with declining propagule supply.

Approaches to setting ballast discharge standards have relied primarily on expert opinion to evaluate the risk–release relationship. The associated history and process (discussed in Chapters 2 and 5 respectively) have resulted in an array of different international, national, and state discharge guidelines and regulations that seek to reduce propagule supply below that of untreated ballast water. These differences result from both uncertainty about the risk–release relationship and from the diverse approaches of different decision makers and stakeholders. Despite uncertainty about the risk–release relationship (discussed further below), the IMO standards are clearly a first step forward, serving to reduce propagule pressure and thus the scale (number and rate) of invasions. In the initial (D-1) IMO phase, the use of open-ocean ballast water exchange (BWE) serves to routinely reduce the concentration of coastal organisms compared to unexchanged water (Ruiz and Reed, 2007; Bailey et al., 2011; Figure 1-6). In a second (D-2) phase, ballast water exchange is to be replaced by treatment technologies to achieve specific discharge standards, in which allowable concentrations of organisms vary by taxonomic group and by size. For zooplankton greater than or equal to 50 μm, the D-2 standard provides a further reduction in coastal organism concentrations beyond that achieved with ballast water exchange (Minton et al., 2005).

Ballast water exchange has been implemented in the U.S. and many other countries (Chapter 2), but its quantitative effectiveness on reducing invasions is not yet known. To our knowledge, no new ballast-mediated invasions have been reported in the Great Lakes or San Francisco Bay in the five year period between 2006 and 2010. This is encouraging and may suggest a decrease of invasions coincident with widespread use of ballast water exchange for these two locations; however, we urge great caution in any interpretation of a causal relationship at the present time. Such conclusions are premature, because of limitations of the short and uneven baseline data. Mandatory ballast water exchange commenced in the United States only in 2007, and even though this began earlier in the Great Lakes, such treatment was not applied to NOBOB vessels until more recently. Importantly, neither region has any coordinated or standardized set of measures designed explicitly to detect new invasions. There has been no formal analysis in the last five years for either location to evaluate time lags between establishment and detection, which can obfuscate reliable measures of changing invasion rates (Ruiz et al., 2000; Solow and Costello, 2004). Moreover, there have been other periods in the past (pre-ballast water exchange) where no new nonindigenous species invasions related to ballast water have been detected in these two systems.

In general, invasion probability increases with propagule pressure when conditions in the recipient system (such as a bay or port) are suitable for colonization. As outlined in Chapter 4, however, the inflection point(s) in this relationship are simply not known. These inflection points are required to predict invasion probability over the operational range of discharge concentrations being considered. It is critical to also recognize that this relationship will vary with the nature of propagule pressure itself (e.g., species richness, abundance, frequency of inoculation, life stage, and genetic composition) as well as source region, recipient region, taxonomic group, season, voyage conditions, and many other potential variables (see Chapter 3). Such variation has important implications for establishing discharge standards, in that invasion probability is obviously context dependent, with many potential influences beyond simply the number of propagules delivered.

Looking forward, there are several approaches that can be used to predict the risk–release relationship, using models described in Chapter 4. One or several of these models should be pursued in the months and years ahead to provide the foundation for a robust, scientific basis for setting and refining specific discharge standards. At the present time, none of the available models have been validated, due mainly to a lack of key data, creating too many untested assumptions to provide confidence in the resulting estimates of invasion probability. Any successful approach also must consider the range of environmental and ecological conditions, geographic locations, and taxonomic groups (with different life cycle strategies, physiological capacities, habitat requirements) encompassed in the global scale of shipping.

The sections below outline current limitations associated with models, including data needs, and a strategy to address these critical data gaps and quantify the risk–release relationship.

MODELS AND DATA GAPS

All of the quantitative models reviewed in Chapter 4 seek to explain or predict the number of invasions (probability of successful establishment) as a function of propagule supply. These approaches assume that (a) reliable measures of propagule supply and invasion outcome are available over some meaningful spatial, temporal, and taxonomic range or (b) this relationship can be reliably derived from first principles (theory) in population biology. At the present time, neither of these criteria is met. There is a critical lack of data needed to populate empirical models or to validate theoretical ones.

To directly estimate the risk–release relationship in any of these approaches, one needs to know (or estimate reliably) the number of viable individuals of each species delivered in ballast water discharge to a particular location (e.g., a port), over some spatial–temporal scale. There are few locations around the globe, let alone the U.S., for which direct measures of propagule supply in ballast water are available. In North America, these include Port Valdez (Alaska), Coos Bay (Oregon), San Francisco Bay (California), Puget Sound (Washington), the Great Lakes, and Chesapeake Bay (see Table 4-2). However, most of these historical ballast data are collected for a very limited time horizon and sample size. Among the larger studies, for example, Carlton and Geller (1993) sampled 157 vessels arriving in Coos Bay from 1986–1991; Hines et al. (2000) sampled 196 vessels arriving in Port Valdez from 1999–2001, and Cordell et al. (2009) sampled 372 vessels arriving in Puget Sound from 2000–2007.

While ballast studies have provided important insights into associated biota and concentrations, it is also critical to recognize that the taxonomic resolution has been insufficient to obtain occurrence and abundance measures for most species. This limitation exists because many ballast water organisms are larvae that cannot be identified to species based simply on morphology. Current molecular tools can permit identification of many but not all species, but these tools have largely been developed since most of the intensive ballast water sampling was conducted in the U.S., and even with genetic techniques quantitative estimates may still be a challenge. Thus, estimates for historical propagule supply from ballast can at best provide a temporal snapshot of concentrations of total organisms and selected taxa at only a few ports.

As noted above, to determine the risk–release relationship using historic information it is also critical to know the number of invasions that have occurred and are attributed to ballast water at the same location and same time frame. There are many retrospective analyses that examine the invasions attributed to ballast water (and other vectors), underscoring the significance of this vector. However, there are also some important limitations in these data. First, the probability of detection has been highly variable in space and time, due to differences in search effort, expertise, and other factors (Ruiz et al., 2000; Ruiz and Hewitt, 2002). Remarkably, there is no current national program designed to provide standardized, field-based measures to detect new invasions and track status and trends (i.e., temporal and spatial patterns) for coastal invasions, despite the implementation of national programs to reduce invasions. Instead, analyses are based largely on "by-catch" data (invasions encountered in the course of other studies) or on localized and infrequent surveys. Thus, the available data result from independent and variable efforts, usually with different aims and methods, limiting inferences that make drawing comparisons through space or time difficult.

Second, there is uncertainty about the vector(s) responsible for many coastal invasions, because multiple vectors are possible for some species. A large percentage of species that arrive in estuarine and marine waters in the past decades is polyvectic, sharing ballast water, hull fouling, sea chest fouling, and other vectors as possible sources (Cohen and Carlton, 1995; Fofonoff et al., 2003; Ruiz et al., 2011). This occurs because species have one or more life stages that can be entrained by multiple vectors. For example, the same species of barnacle may conceivably be transported by a given vessel both in hull fouling as adults and in ballast water as larval stages.

The data on ballast-mediated propagule supply are arguably best for holoplankton, such as calanoid copepods, which have adult forms that can be identified to species, unlike most meroplankton (e.g., Cordell et al., 2009). In addition, holoplankton are less likely to be polyvectic and are attributed primarily to ballast water. However, even for this group it is challenging to estimate a quantitative risk–release relationship because of limitations (i.e., the patchy and incomplete nature) in the measures for propagule supply and/or invasion rate. It is therefore apparent that additional data are needed for directly estimating risk–release relationships using any models, as discussed below.

Proxy Variables

Due to the current lack of direct, species-specific data for determining propagule pressure, some analyses have attempted to use proxy measures to examine risk–release relationships. For example, one might use the number of people (or vehicles) visiting a national park as a proxy for propagule pressure of nonindigenous species to that park and test for a relationship to the number of established nonindigenous species (Lonsdale, 1999; Stohlgren et al., 2003). In many cases, good statistical relationships between variables have been observed, but the practitioner must be careful that the employed proxy is in fact causal and not simply correlative.

In the case of ballast-mediated invasions, several studies have used the number of ship arrivals as a proxy for propagule supply. Even if a relationship is found between vessel arrivals and invasions (in space or time), this relationship may be spurious and not causal. In fact, the number of ship arrivals is a poor proxy for propagule supply, in large part because of high variation in ballast volume discharge among vessel arrivals. Overall, the majority of ships report that they do not discharge ballast water upon arrival (Falkner et al., 2009; Miller et al., 2010), stating that they are offloading cargo, do not have ballast water aboard, or simply do not need to discharge it. In contrast, some vessels can discharge > 50,000 metric tons of ballast upon arrival in the same port. Moreover, there are large differences among vessel types and arrival ports in per-capita ballast discharge (Carlton et al., 1995; Verling et al., 2005; Miller et al., 2010). Together, these factors result in a poor relationship between ballast discharge volume and ship arrivals across ports (see Chapter 1).

Ballast water discharge volume is another proxy variable and an improvement over vessel arrivals, in that it captures one important source of variation among vessels. Combined with information on organism concentrations, this proxy could allow estimates of either total number of propagules per discharge or cumulative propagule supply over some temporal and spatial scale. Nonetheless, volume by itself is unlikely to be representative of propagule supply because significant variation occurs among vessels in the diversity and concentration of biota in their ballast water and sediments. This variation reflects different source locations, seasons, age and management of the water, voyage conditions, and other variables (Smith et al., 1999; Minton et al., 2005; Verling et al., 2005). For example, total zooplankton concentration in ballast water is known to vary five orders of magnitude (Minton et al., 2005). Thus, there is a considerable range of uncertainty in translating ballast water volume to total propagule supply, even before considering species-level information. This may be one reason why a significant relationship between total ballast discharge volume and reported invasions was not observed among 24 estuaries in the U.S. (Lee et al., 2010).

Finally, Lee et al. (2010) attempted to translate ballast volume into total organism (zooplankton) discharge per vessel and estuary, using a pooled distribution of organism concentrations observed for sampled ballast tanks across several discreet locations. While this is conceptually a step forward, ballast volume effectively becomes the key (proxy) variable for propagule supply in the analysis, which cannot yet capture estuary-specific or regional differences that exist in the composition and concentrations of organisms delivered in ballast (as discussed in Chapter 4). This approach represents the limits of current data that exist for ballast biota delivered to most ports (see above).

Clearly, there is potential value in using proxy variables to characterize propagule supply associated with ballast discharge. If reliable proxies could be established, such as the actual volume of ballast water discharged, this would represent an efficient and cost-effective approach to tracking propagule delivery into the future, and it may also open up enormous potential to examine historical data. To date, the performance of available (proposed) proxies has been of limited utility, undermined by the highly variable biotic assemblages of ships operating on a truly global scale (see Chapter 3). Nonetheless, further research into development of possible proxies is warranted, given the potential pay-offs in tracking past, present, and future propagule pressure. Moreover, with the onset of new regulatory and monitoring regimes described below, the ability to use proxies to track future propagule pressure with more precision will be enhanced.

STRATEGIES FOR MOVING FORWARD: GATHERING OBSERVATIONAL AND EXPERIMENTAL DATA

To advance the state of understanding about the risk–release relationship requires the use of one or several modeling approaches (see Chapter 4) and meeting associated data requirements. Because of existing data gaps, past analyses have relied on either proxy variables that have been of limited value or theoretical approaches that have not yet been validated. There are two sources of data that can be used to quantify the risk–release relationship: observational (descriptive) data and experimental data. Both require a substantial commitment in the design, implementation and analysis stages, and sufficient replication to demonstrate broad applicability and generality of results. Generation of usable and meaningful descriptive data will also require a substantial time horizon. Importantly, the same data from either source can be used in a variety of the different models described in Chapter 4, as well as multiple models to test for concordance, as many models share some data requirements. However, the earlier a preferred model is identified, the earlier that data collection can commence to serve the needs of the model most efficiently. The sections below discuss the merits and constraints for each data type.

Descriptive Data

Descriptive data can be collected from analyses of field-based sampling to directly measure propagule supply delivered in ballast water and the associated (detected) rate of invasion. The advantage with this approach is that it could potentially describe the actual relationship that exists in the field. It implicitly includes all of the real-world factors (environmental and biological variables, as well as appropriate spatial and temporal scales) that may affect the outcome of inoculation and the probability of subsequent establishment, whether these are measured or not (Diamond, 1986).

A descriptive approach requires two types of field-based surveys—ship discharge monitoring and receiving system monitoring—to estimate the relationship between propagule delivery and invasion probability for one or several estuaries. A well-designed ship discharge sampling program to measure propagule supply involves stratified random samples across ship type, source regions, and season that is repeated across years. For each ship sampled, the goal would be to characterize the abundance of individual species or taxonomic group of interest, creating a species- or taxon-specific estimate of propagule supply. If the goal is a comprehensive assessment of sampled ballast, including larval forms, this would also require molecular analyses to identify propagules to species. It should be noted that the nature of propagules in ballast tanks is changing, as more and more vessels arrive in the U.S. without intact, original ballast water from a foreign port or estuary, but rather (due to ballast exchange requirements) with largely oceanic (high seas) holo- and meroplankton. Because concentrations of coastal plankton are expected to decrease through time with increasingly stringent ballast management, it is important to consider the effects of temporal changes in evaluation of propagule supply.

In parallel to ballast water analyses, standardized field surveys are needed for the same estuaries, repeated over time, to estimate the occurrence of new invasions for the same taxa or species found in arriving ballast. Ideally, these surveys would involve many taxa across habitats, since it is not possible at the present time to select a subset of "indicator" species that are known to be representative of invasions in general. Analyses of survey results should also consider the occurrence of polyvectic species in evaluating new invasions, to address whether ballast is the probable mode of introduction.

While obtaining such measures of propagule supply and invasion outcome for a single estuary (port system) is necessary for evaluating the risk–release relationship in the field, it is not sufficient to address generality. That is, an assessment in any one estuary is not likely to be representative. The risk–release relationship is expected to differ greatly among locations, both due to differences in propagule supply and also to invasion susceptibility (Lonsdale, 1999). For example, this relationship may be very different among the Great Lakes, San Francisco Bay, Chesapeake Bay, or Tampa Bay (Florida). To achieve a robust understanding, it would be necessary to replicate this descriptive approach in several estuaries, ideally in different geographic regions, considering differences in (a) source region of vessels and associated propagules, (b) recipient region environmental conditions, and (c) recipient region hydrographic regime and retention.

A critical feature and disadvantage of such descriptive measures is the time horizon. Under the best of circumstances in terms of survey intensity, it is estimated to take a decade to (a) estimate propagule supply and (b) detect some of the resultant invasions, accounting for time lags in discovery of nonindigenous species in the field. The nature of this approach is a retrospective analysis, requiring several years to accumulate sufficient data to adequately represent inoculation–establishment outcomes (the relationships between propagule supply and invasion rate).

Experimental Data

Experimental measurements have advantages and disadvantages that are somewhat complementary to those for descriptive measures. With experiments, it is possible to know and control precisely the propagule pressure for individual or multiple species, exposing target laboratory or mesocosm communities under many different conditions, and measure the outcome. As with descriptive measures, it is critical to replicate experiments across a range of taxonomic groups and environmental conditions, to provide robust data. Even with a large number of experiments, the time horizon here to acquire these data is likely to be much shorter than what would be necessary for descriptive data.

The principal limitation with an experimental approach is the potential inability to capture all of the relevant variables, spatial scales, and taxonomic groups. Mesocosms are artificial settings that cannot precisely mimic real-world conditions, despite every effort to do so. For example, the relatively large surface-to-volume ratio in mesocosms may influence plankton behavior and survival in ways not found in nature; the larger the mesocosm, the more one can reduce artifacts associated with walls and differential light penetration and the lower the propagule density of the inoculum that one can achieve, thereby providing greater reality to ballast discharges. In addition, it is not known *a priori* which species or taxonomic groups are most representative for examining invasion dynamics (i.e., the risk–release relationship) in general. Thus, caution is warranted in interpreting the data from such experiments relative to translating the results to real-world scenarios (see Diamond, 1986).

Terrestrial ecologists have long conducted colonization experiments at comparatively large scales (e.g., mangroves studied by Simberloff and Wilson, 1969), and have manipulated propagule pressure and other variables (Vila et al., 2008; MacDougall and Wilson, 2007). In general, less experimental progress has been made in aquatic colonization studies (but see Chadwell and Engelhardt, 2008). Typically either the volume used in experiments is too small and/or the density of colonists too high to realistically simulate the dynamics of ballast release (although see Bailey et al., 2009). Given that large-scale experiments have been used to assess ecological effects of deforestation, lake acidification, and cultural eutrophication (Ricklefs and Miller, 1999), there is no reason to believe that similar studies cannot be conducted to assess risk–release relationships in freshwater or marine communities, using large-scale experimental mesocosms. These would need to either be closed (quarantine) systems or use local biota to prevent any accidental release, but this is not technologically challenging to achieve. Ideally such studies would be conducted with an array of species with different life histories, reproductive modes, sizes, and source regions to represent the range historically and currently observed in ballast discharges to the U.S. Even developing extensive mesocosm facilities necessary for this approach, and with extensive replication, such experiments can be implemented with a fraction of the effort, cost, and time needed for descriptive measures.

While experiments offer a powerful and relatively rapid approach to understanding invasion dynamics, they cannot capture all of the variables and scales that may affect ballast-mediated invasions. One approach to minimize this drawback may be to conduct experiments under "best-case" conditions that are favorable to establishment and that would represent a conservative estimate and a precautionary approach. For example, conditions might simulate a small lagoon or port with high retention, favorable abiotic conditions (e.g., temperature and oxygen), ample food, and low predation or competition. In addition, target species for experiments could also be selected from this perspective, to include a range of taxa with frequent invasions around the globe or with attributes (e.g., parthenogenic, fast-growing, trophic or habitat generalist) considered especially conducive to establishment.

CONCLUSIONS AND RECOMMENDATIONS

The current state of science does not allow a quantitative evaluation of the relative merits of various discharge standards in terms of invasion probability (see the Statement of Task in Chapter 1). Of the approaches suggested to date and reviewed in this report, descriptive statistical modeling with proxy variables (such as the per capita invasion probability approach; see Chapter 4) is currently available to empirically examine the risk–release relationship because there are data available for ballast volume (derived indirectly from vessel arrival data) and historical invasion rates across estuaries. However, it must be cautioned that these are extraordinarily coarse-level data because (a) as noted above, vessel arrival data often do not directly translate into a measure of ballast water actually discharged, (b) when actual ballast volume data are available, these do not translate well into known propagule supply, such as species richness or abundance, and, further, (c) there is no significant relationship between ballast volume and invasions. In addition, the actual number of historical invasions in all estuaries is considered to be underestimated, perhaps significantly so (Carlton, 2009). Thus, while statistical modeling has been applied to current datasets, the data are not sufficient in present form to characterize a biologically meaningful relationship, much less estimate the associated uncertainty, to be able to identify with confidence the invasion probabilities associated with particular discharge standards (propagule supplies).

Several actions are needed to advance a robust understanding of the risk–release relationship in order to inform decisions about ballast water discharge standards. As a logical first step, a benchmark discharge standard should be established that clearly reduces concentrations of coastal organisms below current levels resulting from ballast water exchange (such as the IMO D-2 standard). This will serve to reduce the likelihood of invasion in coastal ecosystems beyond that at the present time.

Following the setting of an initial benchmark, a risk–release model or models should be selected as the foundation for the data gathering and analysis effort. What model or models is ultimately chosen will reflect the available resources, in terms of time, data, and personnel. Indeed, using multiple models with the same data could be valuable to test for concordance. This would also allow one to assess the range of outcomes that would result from proposed ballast water discharge standards. Furthermore, there is considerable worth in transitioning from simple single-species models to more complex multi-species mechanistic models as more data become available.

Finally, a two-track approach should be pursued to obtain both experimental and field-based (descriptive) data. Experiments can be used to evaluate the risk–release relationship and should deliver results over the next three to five years. Field-based descriptive data should also be collected and analyzed to parameterize the same types of models, providing real-world validation of experimental data. Results from such efforts would be expected to materialize in about ten years.

For both experimental and descriptive measures, there is a clear need for a highly directed and coordinated effort. Existing data are limited, and there is no reason to expect a different outcome from future research without a concerted focus. There is a critical need for an explicit sampling design, standardized methods and analyses, and data integration across multiple sites and times. A consistent framework and sufficient oversight and coordination will be necessary to produce the high-quality data specifically needed to populate risk–release models, especially where multiple locations or organizations are involved. Regardless of the structure, any such program must include a centralized data repository, along with active periodic audits on data gathering efforts over time to ensure a standardized methodology is being employed.

There are many potential designs for the structure, implementation, and actual measures that would make up each descriptive and experimental study to quantify the risk–release relationship. Elaboration of the specific designs and protocols is certainly beyond the scope of this report and would require a separate decision making process about particular models to be used and the level of effort that would be pursued. The following recommendations are intended to provide guiding principles for that purpose.

Recommendations for Experiments

Experiments should be used to estimate the effect of propagule pressure on establishment success, using statistical and probabilistic models (Chapter 4). The experiments should (a) be conducted in large-scale mesocosms designed specifically to simulate field conditions, (b) include a diverse range of taxa, encompassing different life-histories and species from known source regions of potential invasions, and (c) include different types of environments (e.g., fresh water, estuarine, and marine water) where ballast discharge may occur.

While it is possible to identify a diverse range of experiments that would be highly informative, initial experimental efforts should focus especially on single-species risk–release relationships. Ideally, these would include taxa and conditions that are selected as "best case" scenarios, seeking to maximize invasion success and provide a conservative estimate of invasion probability (as discussed previously). Thus, rather than experiments that examine complex and interactive effects of many different environmental and biological variables, a premium is placed on relatively simple initial experiments that provide a significant amount of data across "model" taxa and conditions in a short amount of time. This approach should be applied to multiple species, and serious consideration should be given (and explicit criteria used) to select the appropriate organisms and conditions.

The experiments should be advanced aggressively, in a directed fashion, to yield results in a three- to five-year time horizon. While this represents a significant investment in effort and resources, it is the most cost- and time-efficient path to obtaining critical data needed to parameterize risk–release models compared to field-based measures (see below). Experiments could potentially identify a solid interim basis for discharge standards, noting the inherent challenges in working with a limited number of species and the assumptions that these would be representative of a broad array of potential invasions. Importantly, these data may also have direct application to other vectors, in addition to ballast water, as they test basic questions about establishment that are relevant to propagule pressure arising from all vectors.

Recommendations for Descriptive Studies

In addition to experiments, descriptive field-based measures are recommended to ground-truth the models, providing a critical validation step to confirm that (a) risk–release relationships are consistent with experimental results and (b) observed invasion rates are consistent with these predictions. Implementing such an effort at one location is not sufficient. This should occur at selected sentinel estuaries (e.g., San Francisco Bay, Chesapeake Bay, and Tampa Bay), chosen to include different coasts, ship traffic patterns, source regions, and environmental conditions. For each sentinel estuary, measures of propagule supply (in ships' ballast) and invasion rate would be made repeatedly over a minimum of a ten-year time horizon to provide a data set for independent analysis and validation of experimental results.

The specific design of data collection needs to be defined explicitly, considering the model(s) being used and making sure that the output will represent the risk–release relationship and directly translate to a discharge standard. While it may be reasonable to explore potential proxy variables as one component, it is critical to not focus extensively on proxies or other variables that may not represent the risk–release relationship. Also critical is an *a priori* estimate of the uncertainty explicit at all scales, as well as sampling effort (number and frequency of measures), in order to properly design measures and interpret and compare predictions. The same data could be used for statistical and probabilistic models, moving toward increasing resolution (e.g., hierarchical probability models (HPM) as described in Chapter 4) if and as appropriate data are available. However, collecting the data required to examine multiple species across multiple estuaries with HPM would require a large effort, sustained over a considerable period of time. As outlined previously, a comprehensive model would require sampling many vessels (stratified by vessel type, season, and source) and quantifying the concentration of each species present in discharged ballast (as well as volume per discharge event). Field surveys to detect invasions of these species would also need to be conducted coincident with ballast measures.

One possible strategy would be to focus on a subset of target species discharged in ballast water to multiple estuaries. This would reduce considerably the effort required for analysis of ballast water, compared to characterizing the entire community. It may also serve to reduce the sampling effort, and increase the probability of detection, of the target species in field surveys. Intuitively, it would make sense to focus particular attention not only on species that can be identified and counted in ballast samples, but also on species that are unlikely to be polyvectic (such as copepods and mysids), providing the clearest signal (and least noise) for analysis of risk–release relationships associated with ballast water. With this strategy, selection of taxa is critical and should take into consideration biological and environmental requirements (especially whether suitable conditions exist in the specific estuary). As discussed earlier, a challenge is how generally representative any such species would be. Nonetheless, this would result in single-species models (for multiple species) in parallel to the experimental approach outlined previously.

While collection of field-based descriptive data required for a meaningful analysis of the risk–release relationship is somewhat daunting in scope, recent developments make this more feasible than in the past, with continuing rapid improvements in the sensitivity and efficiency of analytical methods. First, pending international and national regulations require commercial vessels to install sampling ports that provide representative and standardized samples of ballast discharge (see Chapter 2); a similar requirement already exists under California state law (California State Lands Commission, 2010). This will provide an important platform for ready access and standardized, comparable samples across vessels and locations. Methods for compliance testing and analyses are already being developed to further standardize data that results from any sampling. Second, the implementation of ballast water treatment systems will reduce the concentrations, and possibly the diversity, of organisms in ballast water (Chapters 1, 3, 4). This may serve to simplify the sampling, having less biological material to process for quantitative analysis. Third, the use of molecular genetic tools has dramatically expanded the capacity (and reduced the time, effort, and cost) to detect species, based on DNA. Several methods already exist that provide high sensitivity in detecting target species in very large volumes of water, including the ability to estimate abundance with quantitative PCR, and the capacity of Next Generation Sequencing is now available for species detection and analyses at the scale of entire communities (Mardis, 2008a,b;

Darling and Mahon, 2011). Importantly, these methods are also rapidly improving in terms of sensitivity, speed, and cost.

Although motivated by understanding the risk–release relationship, it is perhaps useful to point out that field-based measures outlined above would also serve a broader range of applications. These data can provide critical feedback for adaptive management, identifying performance of discharge standards to reduce invasions (see Ruiz and Carlton, 2003). Field-based surveys provide an important baseline, now lacking, to track invasions from other vectors concurrently. In addition, such surveys would align with recommendations from the Interagency Ocean Policy Task Force (2010). The Task Force plan subdivides the United States into nine regional planning areas and recommends the federal government support "disciplinary and interdisciplinary science, research, monitoring, mapping, modeling, forecasting, exploration, and assessment to continually improve understanding of ocean, coastal, and Great Lakes ecosystems" (Interagency Ocean Policy Task Force, 2010).

***

To date, there has been no concerted effort to collect and integrate the data necessary to provide a robust analysis of the risk–release relationship needed to evaluate invasion probability associated with particular ballast water discharge standards. Existing experimental and field data are of very limited scope. There is currently no program in place to implement either ship-based ballast sampling or field surveys to detect new invasions across sites. On-going research provides confidence that this approach is feasible, but it is scattered across sites and usually short-term in nature. Several models exist which can quantify the risk–release relationship, given sufficient data that are now lacking. This report outlines the paths, using multiple methods over different time frames, that could address these data gaps, and thus provide a robust foundation for framing scientifically supportable discharge standards for ballast water.

REFERENCES

Bailey, S. A., L. A. Velez-Espino, O. E. Johannsson, M. A. Koops, and C. J. Wiley. 2009. Estimating establishment probabilities of Cladocera introduced at low density: an evaluation of the proposed ballast water discharge standards. Canadian Journal of Fisheries and Aquatic Sciences 66:261–276.

Bailey, S. A., M. G. Deneau, L. Jean, C. J. Wiley, B. Leung, and H. J. MacIsaac. 2011. Evaluating efficacy of an environmental policy to prevent biological invasions. Environmental Science and Technology 45:2554–2561.

Carlton, J., and J. Geller. 1993. Ecological roulette: The global transport of nonindigenous marine organisms. Science 261:78–82.

Carlton, J. T., D. M. Reid, and H. van Leeuwen. 1995. Shipping Study. The role of shipping in the introduction of non-indigenous aquatic organisms to the coastal waters of the United States (other than the Great Lakes) and an analysis of control options. The National Sea Grant College Program/Connecticut Sea Grant Project R/ES-6. Department of Transportation, United States Coast Guard, Washington, DC and Groton, Connecticut. Report Number CG-D-11-95. Government Accession Number AD-A294809.

Carlton, J. T. 2009. Deep invasion ecology and the assembly of communities in historical time. Pp. 13–56 In: Biological Invasions in Marine Ecosystems. G. Rilov and J. A. Crooks, editors. Berlin: Springer-Verlag.

Chadwell, T. B., and K. A. M. Engelhardt. 2008. Effects of pre-existing submerged vegetation and propagule pressure on the invasion success of *Hydrilla verticillata*. Journal of Applied Ecology 45:515–523.

Cohen, A. N., and J. T. Carlton. 1995. Biological Study. Nonindigenous Aquatic Species in a United States Estuary: A Case Study of the Biological Invasions of the San Francisco Bay and Delta. A Report for the United States Fish and Wildlife Service, Washington, DC, and The National Sea Grant College Program, Connecticut Sea Grant, NTIS Report Number PB96-166525.

Cordell J., D. Lawrence, N. Ferm, L. Tear, S. Smith, and R. Herwig. 2009. Factors influencing densities of non-indigenous species in the ballast water of ships arriving at ports in Puget Sound, Washington, United States. Aquatic Conservation: Marine and Freshwater Ecosystems 19:322–343.

Darling, J. A., and A. R. Mahon. 2011. From molecules to management: Adopting DNA-based methods for monitoring biological invasions in aquatic environments. Environmental Research doi:10.1016/j.envres.2011.02.001.

Diamond, J. 1986. Overview: laboratory experiments, field experiments, and natural experiments. Pp 3–23 In: Community Ecology. J. Diamond, and T. J. Case, editors. New York: Harper and Row.

Fofonoff, P. W., G. M. Ruiz, B. Steves, and J. Carlton. 2003. Ships or On Ships? Mechanisms of transfer and invasion for non-native species to the coasts of North America. Pp. 152–182 In: Invasive Species: Vectors and Management Strategies. G. M. Ruiz, and J. T. Carlton, editors. Washington, DC: Island Press.

Hines, A. H., G. M. Ruiz, and L. S. Godwin. 2000. Assessing the risk of non-indigenous species invasion in a high latitude ecosystem: ballast water treatment facility in Port Valdez, Alaska. Pp. 81–88 In: Marine Bioinvasions, Proceeding of the First National Conference, MIT Sea Grant College Program, Cambridge. J. Pederson, editor. Cambridge, MA: Massachusetts Institute of Technology.

Interagency Ocean Policy Task Force. 2010. Final recommendations of the Interagency Ocean Policy Task Force. The White House Council on Environmental Quality, Washington, DC.

Lonsdale, W. M. 1999. Global patterns of plant invasions and the concept of invasibility. Ecology 80:1522–1536.

MacDougall, A. S., and S. D. Wilson. 2007. Herbivory limits recruitment in an old-field seed addition experiment. Ecology 88:1105–1111.

Mardis, R. E. 2008a. Next-Generation DNA Sequencing Methods. Annual Review of Genomics and Human Genetics 9:387–402.

Mardis, R. E. 2008b. The impact of next-generation sequencing technology on genetics. Trends in Genetics 24(3):133–141.

Miller, A. W., T. Huber T, M. S. Minton, and G. M. Ruiz. 2010. Status and trends of ballast water management in the United States: Third Biennial Report to the United States Coast Guard. Edgewater, MD: National Ballast Information Clearinghouse, Smithsonian Environmental Research Center (SERC).

Minton, M., E. Verling, A. Miller, and G. Ruiz. 2005. Reducing propagule supply and coastal invasions via ships: effects of emerging strategies. Frontiers in Ecology and the Environment 3:304–308.

Reusser, D. 2010. Biogeography of Nonindigenous Species: From Description to Prediction. Corvallis, OR: Oregon State University.

Ricklefs, R. E., and G. Miller. 1999. Ecology. Fourth edition. San Francisco, CA: W. H. Freeman.

Ruiz, G. M., and J. T. Carlton. 2003. Invasion vectors: a conceptual framework for management. Pp. 459–504 In: Invasive Species: Vectors and Management Strategies. G. M. Ruiz, and J. T. Carlton, editors. Washington, DC: Island Press.

Ruiz, G. M., and C. L. Hewitt. 2002. Toward understanding patters of coastal marine invasions: A prospectus. Pp. 529–547 In: Invasive aquatic species of Europe. E. Leppakoski, S. Olenin, and S. Gollasch, editors. Dordrecht: Kluwer Academic Publishers.

Ruiz, G. M., and D. F. Reid (editors). 2007. Current state of understanding about the effectiveness of ballast water exchange (BWE) in reducing aquatic nonindigenous species (ANS) introductions to the Great Lakes Basin and Chesapeake Bay, USA: Synthesis and analysis of existing information. NOAA Technical Memorandum GLERL-142, National Ocean and Atmospheric Administration, Ann Arbor.

Ruiz, G. M., P. Fofonoff, J. T. Carlton, M. J. Wonham, and A. H. Hines. 2000. Invasions of Coastal Marine Communities in North America: Apparent Patterns, Processes, and Biases. Ann. Rev. Ecol. Syst. 31:481–531.

Ruiz, G. M., P. W. Fofonoff, B. Steves, S. F. Foss, and S. N. Shiba. 2011. Marine invasion history and vector analysis of California: A hotspot for western North America. Diversity and Distributions. 17: 362–273.

Smith, L. D., M. Wonham, L. McCann, G. Ruiz, A. Hines, and J. Carlton. 1999. Invasion pressure to a ballast-flooded estuary and an assessment of inoculant survival. Biological Invasions 1:67–87.

Solow, A., and C. Costello. 2004. Estimating the rate of species introductions from the discovery record. Ecology 85:1822–1825.

Stohlgren, T. J., D. Barnett, and J. Kartesz. 2003. The rich get richer: pattern so plant invasions in the United States. Frontiers in Ecology and Environment 1:11–14.

Verling, E., G. M. Ruiz, L. D. Smith, B. Galil, A. W. Miller, and K. Murphy. 2005. Supply-side invasion ecology: characterizing propagule pressure in coastal ecosystems. Roy. Soc London B 272:1249–1256.

Vila, M., A. S. D. Siamantziour, G. Brundu, I. Camarda, P. Lambdon, F. Medail, E. Moragues, C. Sueh, A. Traveset, A. Y. Troumbis, and P. E. Hulme. 2008. Widespread resistance of Mediterranean island ecosystems to the establishment of three alien species. Diversity and Distributions 14:839–851.

Glossary

See Carlton and Ruiz (2005) for a framework for the standardization of terms and definitions associated with vector science.

Allee effects — density-dependent phenomena that can cause the per-capita rate of population growth to decline with decreasing population size (Kramer and Drake, 2010)

ballast water exchange — transfer of water from a ship's ballast systems to the environment, with concomitant ("continuous exchange" method) or subsequent ("empty-refill" method) uptake of water; generally intended to take place in off-shore (open ocean) waters (cf. salt-water flushing) with a goal or removing or reducing the original contents and associated biota from coastal or nearshore habitats. Note that the target for exchange (100 percent for empty refill or 300 percent flow-through by volume) is intended to achieve 95 percent replacement of the original water.

ballast water management — (1) in the context of reducing invasions, ballasting practices designed to minimize the uptake or discharge of living organisms; techniques or treatments designed to remove, kill or inactivate prior to their discharge, organisms inadvertently taken up when water is brought into tanks or ballasted holds; (2) independent of invasion biology considerations, the movement of water into and out of ships' ballast tanks and ballasted holds in such a way as to maximize stability and safety of the ship, its crew, and cargo.

cryptogenic — a species that cannot be reliably demonstrated as native or non-indigenous.

diapausing — describes a period during which physiological activity is suspended.

discharge standard — criteria for the maximum concentration of living organisms that may be discharged by a ship when discharging ballast; the value is established by a regulatory or governing body.

donor region — dispersal hub within which a species interfaces with a transport mechanism (Carlton, 1996); synonymous with "source region."

establishment probability — used interchangeably herein with "invasion risk"; the chance that one or more individuals establishes a self-maintaining population.

euryphagic — a species exhibiting broad feeding range.

eurytopic — a species exhibiting broad habitat and physiological breadth.

founder effect — a reduction in genetic diversity occurring when a novel population is established by a relatively small number of colonists.

holoplankton — organisms having their entire life history in the planktonic (free-swimming) state.

inoculum density (ID or D_I) —the total number of organisms in the inoculum (N_I) divided by the volume of the inoculum (V_I) (Minton et al., 2005), i.e., $D_I = N_I / V_I$.

initial population size — the initial number of organisms released into the environment in a given location at a given time, i.e., the inoculum number (N_I). In its simplest formulation, the organisms' density in the environment (D_E) is given as the number of organisms released (N_I) divided by the volume of water in the environment (V_E), i.e., $D_E = N_I / V_E$.

invasion rate — the number of nonindigenous species that establish in a given region per unit time.

invasion risk — *see* establishment probability.

invasive species — a broad term for a nonindigenous species to which is attributed environmental, ecological, economic, or societal harm, or a combination of these.

lecithotrophic — refers to planktonic larvae whose nutrition is yolk derived from their egg.

meroplankton — defined as organisms having planktonic life stages for part of their life history.

metapopulations — spatially separated populations of the same species that are linked by dispersal.

microorganism — a wide variety of microscopic organisms consisting of a single cell or cell cluster; includes viruses, which are not cellular (Madigan and Martinko, 2006); synonymous with "microbe."

models — conceptual, graphical, or mathematical descriptions of a phenomenon or process. Here we define two major categories that are not mutually exclusive, but which represent very different modeling philosophies and goals: *descriptive models* represent the shape of the relationship, often in graphic form; *mechanistic models* define the processes generating the relationship (e.g., Drake and Jerde, 2009) and consist of two general classes, probabilistic statements and dynamic, demographic models.

nonindigenous species — a species not naturally (historically) found in a particular geographic location or region; synonymous with non-native, alien, introduced, exotic, and foreign (cf. invasive species).

parthenogenetic — describing a form of reproduction in which an unfertilized egg develops into a new individual, occurring commonly among insects and certain other arthropods.

pathway — a term used to encompass (a) why a species is transported, whether accidentally or deliberately (*cause)*, (b) the geographic path over which a species is transported (*route)* and (c) how a species is transported (*vector*) (Carlton and Ruiz, 2005).

polyvectic — species transported by more than one vector (Carlton and Ruiz, 2005); cf. vector.

probability of establishment (P_E) — probability of establishing a self-sustaining population.

propagule — any living biological material (particles, cells, spores, eggs, larvae, and mature organisms) transported from one location to another.

propagule pressure — a general term expressing the quantity, quality, and frequency with which propagules are introduced to a given location. Propagule pressure is a function of a suite of variables reflecting the nature of the species and the transport vector.

recipient region — endpoint of dispersal spoke at which a species is released (initial point of inoculation) (Carlton, 1996); synonymous with target region.

risk–release relationship — the relationship between invasion risk and propagule pressure for a single or multiple species. It expresses the number of species that establish as a function of the large-scale release of a varying number of varying species at varying densities. The relationship between invasion risk and propagule pressure is generally held to be positive although the shape of the curve may vary considerably.

salt-water flushing — the input of salt water to a ship's seemingly empty or partially filled ballast tank with the intent of (1) imposing a lethal physiological shock to organisms therein or (2) removing residual living organisms and sediments, which are then pumped back out into the open sea ("rinse and spit") (cf. ballast water exchange).

vector — how a species is transported, that is, the physical means or agent (such as ballast, hulls, or movement of commercial oysters) (Carlton and Ruiz, 2005) (cf. polyvectic).

vector strength — the relative number or rate of established invasions that result within a specified time period from a given vector in a particular region (Carlton and Ruiz, 2005).

virus — A genetic element containing either DNA or RNA that replicates in cells but is characterized by having an extracellular state (Madigan and Martinko, 2006).

REFERENCES

Carlton, J. T., and G. M. Ruiz. 2005. Vector science and integrated vector management in bioinvasions ecology: conceptual frameworks. Pp. 36–58 In: Invasive Alien Species. A New Synthesis. Mooney H. A., Mack R. N., McNeely J. A. , Neville L. E. , Schei P. J., Waage J. K., editors. Scope 63. Washington, DC: Island Press.

Carlton, J. T. 1996. Pattern, process, and prediction in marine invasion ecology. Biological Conservation 78:97–106.

Drake, J. M., and C. Jerde. 2009. Stochastic models of propagule pressure and establishment. Chapter 5 In: Bioeconomics of invasive species: integrating ecology, economics, policy, and management. Keller, R. P., Lodge, D. M, Lewis, M. A., and Shogren, J. F. (editors). New York: Oxford University Press.

Kramer, A. M., and J. M. Drake. 2010. Experimental demonstration of population extinction due to a predator-driven Allee effect. Journal of Animal Ecology 79:1365–2656.

Madigan, M. T., and J. M. Martinko. 2006. Biology of Microorganisms. Upper Saddle River, NJ: Pearson Prentice Hall.

Minton, M. S., E. Verling, A. W. Miller, and G. M. Ruiz. 2005. Reducing propagule supply and coastal invasions via ships: Effects of emerging strategies. Frontiers in Ecology and the Environment 3:304–308.

Appendix A

Committee Biographical Information

JAMES T. CARLTON, Chair, is a professor of Marine Sciences at Williams College. He has directed the Williams College/Mystic Seaport Program since 1989 and also teaches marine ecology. He holds a B.A. in paleontology from the University of California, Berkeley, a Ph.D. in ecology from the University of California, Davis, and he was a postdoctoral scholar at the Woods Hole Oceanographic Institution. His research is on global marine bioinvasions (their ecosystem impacts, dispersal mechanisms, and management strategies) and on marine extinctions in modern times. He was the first scientist to receive the federal government's Interagency Recognition Award for his national and international work to reduce the impacts of nonindigenous invasions in the sea. Dr. Carlton has served on two NRC committees including the 1996 Committee on Ships' Ballast Operations, which produced *Stemming the Tide: Controlling Introductions of Nonindigenous Species by Ships' Ballast Water*, and he was co-chair of the Marine Biodiversity Committee, which produced *Understanding Marine Biodiversity: A Research Agenda for the Nation.*

GREGORY M. RUIZ, Vice-Chair, is a marine ecologist and senior scientist at the Smithsonian Environmental Research Center. He is one of the world's foremost experts on invasion biology in coastal marine ecosystems. He directs the National Ballast Information Clearinghouse, the principal aims of which are to quantify the amounts and origins of ballast water discharged in U.S. coastal systems and to determine the likelihood of ballast-mediated invasions by nonindigenous species. Dr. Ruiz also studies the biology, ecology, and patterns of transfer mechanisms (vectors) that deliver species beyond their geographic range; the biological and ecological attributes of species in their non-native range; and global patterns of biological invasions and factors that control observed distributions. He has been involved in numerous national and international working groups dealing with nonnative species invasions, including the

Scientific and Technical Advisory Committee of the EPA's Chesapeake Bay Program and the Global Invasive Species Program. Dr. Ruiz received his B.A. in aquatic biology from University of California, Santa Barbara, and his Ph.D. in zoology from the University of California, Berkeley.

JAMES E. BYERS is an associate professor in the Odum School of Ecology at the University of Georgia. He was previously a professor at the University of New Hampshire, and he has been a visiting scholar at the University of Wollongong and the University of Technology Sydney, both in Australia. Dr. Byers studies species interactions in nearshore, estuarine, and marsh environments, focusing on quantitatively measuring the impacts of nonindigenous species on native biota in invaded communities. He has developed quantitative tools to better understand how and when invading species will impact native systems and to help predict outcomes of future invasions. He has recently begun working on the spread of invasions in advective environments. He received his B.S. in zoology from Duke University and his Ph.D. in ecology from the University of California, Santa Barbara.

ALLEGRA CANGELOSI is the director of Environmental Projects for the Northeast-Midwest Institute. She has been engaged in policy, technological, and scientific aspects of ship-mediated invasive species introductions for two decades, including activities at the state, regional, federal, and international levels. In the 1990s, she directly assisted in fashioning and gaining enactment of the Nonindigenous Aquatic Nuisance Prevention and Control Act and its amendments. From 1996 to 2002, Ms. Cangelosi participated in the U.S. delegation to the International Maritime Organization (IMO) as a technical advisor for the Ballast Water Working Group. During the same period, she led a multidisciplinary team in undertaking ballast treatment performance assessments in operating ship and shore-based contexts, including performance evaluations with respect to zooplankton, phytoplankton, and microbial organisms. Most recently she has developed harbor and ship discharge monitoring methods for detecting species risks, as well as ballast treatment verification in fresh water environments, both via the Great Ships Initiative. She received her B.A. in biology from Kalamazoo College and her M.S. in resource economics from Michigan State University.

FRED C. DOBBS is a professor and the Graduate Program Director in the Department of Ocean, Earth and Atmospheric Sciences at Old Dominion University. He received his A.B. in biology from Franklin and Marshall College, his M.S. in zoology from University of Connecticut, and his Ph.D. in oceanography from Florida State University. His current research addresses several areas in aquatic microbial ecology. In the context of ballast water quality standards, he is known for his interests in the abundance, population dynamics, and survival of microorganisms, including potential pathogens, in ballast tanks. In addition, he has considered the fate of microorganisms discharged into harbor waters and their potential for subsequent growth and establishment. Dr. Dobbs was a charter member and continues to serve on the EPA/ Environmental Technology Verification Ballast Water Technology Panel that considers design and testing of shore-based ballast water treatment systems. He also sits on the Advisory Panel of the Marine Invasive Species Program, California State Lands Commission, and is an Advisory Board Member of the Maritime Environmental Resource Center, University of Maryland.

EDWIN D. GROSHOLZ is a professor and the Alexander and Elizabeth Swantz Specialist in Cooperative Extension at the University of California, Davis. His work focuses on understanding the impacts of introduced species in coastal ecosystems and the relative importance of vectors such as ballast water in their establishment. He also studies the consequences of introduced parasites and diseases in marine and estuarine systems. His involvement with invasive species and ballast water management includes being co-chair of the Coastal Committee for the Western Regional Panel of the Aquatic Nuisance Species Task Force, a member of the Southern California Caulerpa Action Team, co-author of the federal management plan for the European Green Crab, and co-author of the California Aquatic Invasive Species Management Plan. He co-authored the Pacific State Marine Fisheries Commission workshop proceedings on Alternate Ballast Water Exchange Areas and developed a comprehensive invasive species database for the San Francisco Bay-Delta Region that focuses on the origin and likelihood ballast water transport. Dr. Grosholz manages an active outreach and education program focused on preventing invasions of aquatic ecosystems. He received his B.A. in biology from Brown University and his Ph.D. in zoology from the University of California at Berkeley.

BRIAN LEUNG is an assistant professor in the Biology Department at the McGill School of the Environment. The broad underlying theme of his research is the application of ecology to important environmental issues, merging mathematical, computational, and statistical modeling techniques with empirical information. He is currently focusing on invasive species, forecasting when and where biological invasions will occur, what effect they will have, and determining what should be done about them. He uses tools such as bioeconomics, computational biology, and risk assessment as vehicles for decision-making. His research interests also include infectious diseases. From 2001-2002, he was the coordinator of the NSF Biocomplexity Interdisciplinary Project, which evaluated biological and economic risks posed to Great Lakes ecosystems by invading species. He received his B.Sc. in biology from the University of British Columbia, Vancouver, and his Ph.D. in biology from Carleton University, Ottawa, Ontario.

HUGH J. MACISAAC is a professor and the Invasive Species Research Chair in the Department of Fisheries and Oceans at the Great Lakes Institute for Environmental Research, University of Windsor. He is also the director of the Canadian Aquatic Invasive Species Network, a consortium of 34 professors from across Canada who study aquatic invasive species in a systematic manner. Dr. MacIsaac received his M.Sc. from the University of Toronto and his Ph.D. from Dartmouth College. His research has focused on vectors of invasion in freshwater systems, most notably shipping. He is interested in hull fouling as a mechanism of non-native species introduction, ballast water treatment, invasion theory, and the genetic structure of invading species. He has conducted ballast water and hull fouling studies on the Great Lakes, in Halifax, and in Vancouver, and is presently initiating similar work in the Arctic. Dr. MacIsaac was a member of both NRC committees on the St. Lawrence Seaway: Options to Eliminate Introduction of Nonindigenous Species into the Great Lakes.

MARJORIE J. WONHAM is a professor of Life Sciences at Quest University Canada. Her invasion research has included empirical and mathematical approaches to studying ballast water transport, marine invasion success and impacts, and epidemiological invasion dynamics. She has contributed to both national and international invasion biology working groups, including the National Biological Invasions Shipping Study, Impacts of Invasive Species: Toward a Theoretical Framework, the Canadian Aquatic Invasive Species Network, and the Integrated Systems for Invasive Species project. Dr. Wonham also teaches summer courses in invertebrate zoology, conservation biology, and science communciation at Friday Harbor Laboratories, Bamfield Marine Sciences Centre, and the Wrigley Marine Science Center. She holds a B.A. in Natural Sciences from Cambridge University and a Ph.D. in zoology from the University of Washington.

LAURA J. EHLERS is a senior staff officer for the Water Science and Technology Board of the National Research Council. Since joining the NRC in 1997, she has served as the study director for fifteen committees, including the Committee to Review the New York City Watershed Management Strategy, the Committee on Bioavailability of Contaminants in Soils and Sediment, the Committee on Assessment of Water Resources Research, and the Committee on Reducing Stormwater Discharge Contributions to Water Pollution. Ehlers has periodically consulted for EPA's Office of Research Development regarding their water quality research programs. She received her B.S. from the California Institute of Technology, majoring in biology and engineering and applied science. She earned both an M.S.E. and a Ph.D. in environmental engineering at the Johns Hopkins University.

ELLEN DE GUZMAN is a senior program associate for the Water Science and Technology Board of the National Research Council. In addition to working on study projects, she manages the Board's websites and coedits its biennial report. She received her B.A. from the University of the Philippines and her M.A. in international development from American University.